Destruction of Nuclear Energy Facilities in War

Written under the auspices of the Center for International and Strategic Affairs, University of California, Los Angeles, and the Center of International Studies, Princeton University

A list of other Center publications appears at the back of this book

Destruction of Nuclear Energy Facilities in War

The Problem and the Implications

Bennett Ramberg
University of California
Los Angeles

LexingtonBooks
D.C. Heath and Company
Lexington, Massachusetts
Toronto

Library of Congress Cataloging in Publication Data

Ramberg, Bennett.
  Destruction of nuclear energy facilities in war.

  "Written under the auspices of the Center for International and Strategic
Affairs, University of California, Los Angeles, and the Center of
International Studies, Princeton University."
  Includes bibliographical references and index.
  1. War damage, Industrial. 2. Nuclear facilities—Military aspects.
3. Nuclear facilities—Defense measures. I. California. University. University
at Los Angeles. Center for International and Strategic Affairs. II. Princeton
University. Center of International Studies. III. Title.
UA929.95.A87R35   355'.028   80-7691
ISBN 0-669-03767-2

Published simultaneously in Canada.

Printed in the United States of America.

International Standard Book Number: 0-669-03767-2

Library of Congress Catalog Card Number: 80-7691

299425

Contents

                   Nuclear Energy Facilities in War?   161

                   Appendix   165

                   Notes   167

                   Index   189

                   About the Author   195

                   Center of International
                   Studies: List of Publications   197

                   Center for International
                   and Strategic Affairs:
                   List of Publications   203

List of Figures

List of Tables

Preface

This book addresses the issues of why nuclear energy facilities are attractive targets in war, why nations should be concerned about this, and what can be done to minimize risks. It is directed at three audiences. The first includes policy makers and analysts with international strategic concerns and others who plan, regulate, and review nuclear energy development. I hope this work will familiarize them with important issues that heretofore have received little attention. I also have been conscious of the concerns of the nuclear engineers and physicists who design atomic installations. When I began this work, members of the technical community advised me that I would have to demonstrate why systems designed to prevent major accidental releases of radioactivity would not function as a result of wartime bombardment and why the consequences would create significant strategic costs beyond those resulting from other military actions. To meet this challenge, I draw from available literature on accidents to indicate nuclear installation vulnerability to willful destruction and the associated circumstances that will result in serious consequences. The final audience is the interested lay public. In recent years the debate over nuclear energy has captured the attention of the general populace in many countries. Keeping in mind that this audience, as well as some members of the policy community, may not be well versed in nuclear technology, I include elementary technical information necessary to follow the argument.

Acknowledgments

I began this manuscript at the Center of International Studies, Princeton University, and concluded it at the Center for International and Strategic Affairs, University of California, Los Angeles. I am indebted to both institutions and their respective directors, Cyril Black and Roman Kolkowicz, for support. During my research and writing a number of people kindly offered helpful comments. Drafts of the entire study were read by Louis René Beres, Purdue University; Jan Beyea, Princeton University; Conrad Chester, Oak Ridge National Laboratory; Dean Kaul, Science Applications Inc.; Joseph Loftus, formerly of the Rand Corporation; Milton Plesset, California Institute of Technology; George Quester, Cornell University; and Bart Sokolow, UCLA. Others who shared their thoughts with me on specific chapters include Cyril Black, Harold Feiveson, George Luchak, Theodore Taylor, and Frank von Hippel, Princeton University; Eugene Cramer, Southern California Edison; Kreszentia Duer, World Bank; Gloria Duffy, Arms Control Association; Gunther Handl, University of Texas; and Thomas Ilgen, Brandeis University. Lillian Katz edited the work. I also benefited from the encouragement of many interested individuals. In addition to those listed above I wish to thank Anthony Alperin, Vernon Aspaturian, Lester Chagi, Richard Goldin, Constance Ramberg, Max Ramberg, Ida Siegel, Herbert Siegel, Phyllis Scadron, Richard Scardon, and Jiri Valenta. Above all, my deepest gratitude goes to my parents. Any errors of fact or judgment are mine alone.

Introduction

Despite the likely multiplication of nuclear power plants and their support facilities throughout the world, little public consideration has been given to their vulnerability in time of war.[1] Instead attention has been focused on costs, waste disposal, accidents, plutonium diversion, and sabotage. This situation may be understandable in the United States where war is likely to involve nuclear weapons irradiation of large portions of the country. However, in other regions of the world—Europe, the Middle East, Korea, China, Taiwan, South Asia, West Asia, and southern Africa—where nuclear energy installations are in place or planned, their presence affords combatants a radiological weapon where warfare would otherwise be conventional. The failure or unwillingness of policy makers in the United States and abroad to make this matter a subject for extensive public review and debate is unfortunate.

In one of the few treatments, *Nuclear Power and the Environment*, Great Britain's Royal Commission on Environmental Pollution laid out the dimensions of the problem:

> We have given some thought to the possible effects of war so far as nuclear installations are concerned; these installations, providing vital energy supplies, would be prime targets. In a nuclear war the effects of attack on nuclear installations would be one part of the general catastrophe, but an attack with conventional weapons leading to the release of radioactivity would produce some of the effects of nuclear weapons. The quantities of fission products that could be released are vast and they would not be carried up into the stratosphere. The effects of war, even of "conventional" war, are inevitably horrifying, but if these effects could be magnified by attack on nuclear installations, then this is a major factor to consider whether, or to what extent to use nuclear power. This threat also exists, and should likewise be weighed, in the non-nuclear field. The vast increase in the chemical process industry over the last few decades has created many industrial plants where the consequences of damage from armed attack could be extremely serious. The unique aspect of nuclear installations is that the effects of the radioactive contamination that could be caused are so long lasting. If nuclear power could have been developed earlier, and had it been in widespread use at the time of the last war, it is likely that some areas of central Europe would still be uninhabitable because of ground contamination by caesium.[2]

History supports the Royal Commission's concerns. To cripple an antagonist's industrial ability to wage war, combatants have attacked enemy energy sources during World War II and the Korean and Vietnam wars.[3] Destruction of the environment for military purposes also has precedents. The Dutch destroyed their dikes during World War II in order to hamper the Germans; the Soviet Union followed a scorched-earth policy for the same reason; and during the Vietnam war, the United States employed herbicides to destroy enemy defensive cover and to improve target identification.[4] Nuclear facilities may be attractive targets for other reasons. They represent one of the greatest concentrations of capital investment a country is likely to possess. Their destruction could significantly augment nuclear weapons fallout and thereby impede postwar recovery. A party with a stake in an ongoing conflict between two countries might consider sabotaging a facility to escalate the conflict. Finally many people in many countries have become acutely concerned about the possible release of radionuclides from power plants. Taking advantage of this fear, a belligerent could use the threat of radioactive contamination as a means of coercion.[5]

For these reasons, combatants are likely to contemplate the destruction of atomic installations, including nuclear fuel fabrication, power, spent fuel, reprocessing, and waste storage facilities. Radionuclide discharge would damage the environment significantly. Major accident scenarios, comparable to releases resulting from conventional explosives, indicate that discharges from large depositories could contaminate thousands of square miles. There would be significant implications for international stability should adversaries release or threaten to release radioactive products for purposes of intimidation. However, there are measures also available to minimize dangers.

The purposes of this study are to demonstrate that acts of war should be included in calculations of the risks of nuclear energy in many regions of the world and to suggest means by which such risks can be reduced. The argument is developed in five chapters. Chapter 1 explores the biological impact of radionuclides contained in nuclear energy installations. Chapter 2 reviews the vulnerability of these facilities to wartime destruction and the resulting contamination. Chapter 3 assesses the implications for deterrence, coercive diplomacy, and military strategy in Europe, the Middle East, Asia, Africa, and the United States. Chapter 4 proposes measures to minimize resulting strategic instability through international law, defense, engineering, and energy alternatives that make installations more war resistant and less threatening to public health, and means to strengthen international institutions to monitor the wartime implications of nuclear exports. Chapter 5 weighs the merits of ignoring nuclear facility vulnerability in war.

Destruction of Nuclear Energy Facilities in War

1

The Biological Problem Posed by Nuclear Energy

The biological problem posed by nuclear energy derives from the impact of radiation on living matter.[1] Radiation exists at most stages of the nuclear fuel cycle. Schematized in figure 1-1 for one type of reactor, the cycle involves conversion of uranium ores into fuel that is spent generating energy at the power plant. When the cycle is closed (it is not in most countries at the present time), chemical reprocessing plants extract recyclable plutonium and uranium from spent fuel and send wastes to repositories.

Of greatest environmental concern are the reactor, spent-fuel storage, reprocessing plants, and plutonium and high-level waste repositories. Each of these involves actinides and fission products, highly radioactive elements derived from the fission of fuel in the reactor. The remainder of the cycle contains materials having relatively low radiation.

Actinides and fission products emit four types of biologically significant radiation: alpha particles, beta particles, gamma radiation, and neutrons. Upon penetration of living matter, they transfer energy to cellular material, damaging molecules by breaking chemical bonds and displacing electrons (ionization), which induces further chemical changes. Either or both somatic and genetic consequences may result.

The amount of damage will vary in proportion to the rate of linear energy transfer (LET). Alpha particles and neutrons through proton ionization expend energy quickly, resulting in heavy local damage that, per unit of energy, is greater than that caused by gamma and beta particles, which give up their energy much more slowly but consequently penetrate more deeply. To account for different effects of radiation, several measurements have been divised. The roentgen is a unit of X and gamma ray intensity. The rad reflects the radiation energy absorbed by a material. Roentgen equivalent man (rem) is a measure of the variable effectiveness of any radiation for producing biological consequences. Thus 1 rad of beta and gamma radiation is approximately equivalent to 1 rem; 1 rad of neutron radiation, 4 to 10 rem; and 1 rad of alpha particles, 10 to 20 rem. Since radiation may be accumulated over time, dose rates are measured in time units (for example, 5 rem in one year).[2]

Among hundreds of actinides and fission products contained in irradiated fuel and waste, iodine-131 (^{131}I), strontium-90 (^{90}Sr), cesium-137 (^{137}Cs), and plutonium-239 (^{239}Pu) are probably the most conspicuously deleterious to health. Each affects a different part of the human body and is

1

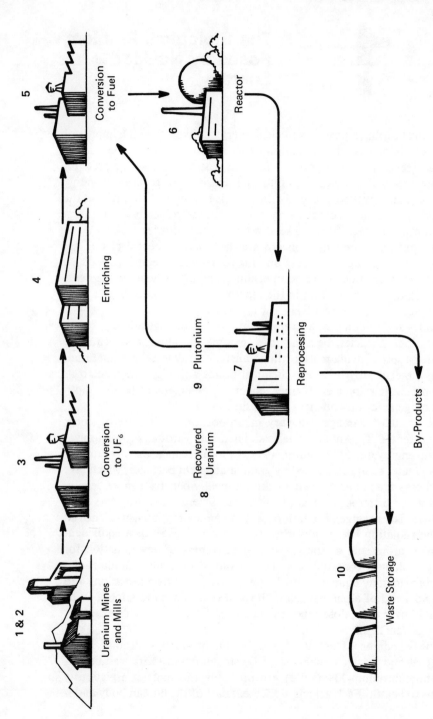

Source: U.S. Atomic Energy Commission, *The Safety of Nuclear Power Reactors and Related Facilities,* WASH 1250 (Washington, D.C.: U.S. Atomic Energy Commission, 1973), p. 4-2.

Figure 1-1. Light Water Reactor Fuel Cycle: Uranium and Plutonium Recycle

retained for varying lengths of time, depending on naturally occurring decay (measured in terms of the product's half-life) and the body's processes of elimination. The length of time it takes the body to remove half of the substance is called the effective half-life of the radioactive product. These half-lives, varying from seven days to almost two hundred years, are shown in table 1-1.

Human somatic and genetic disorders resulting from exposure to these substances vary with the radionuclide, the intensity and duration of exposure, and the age and sex of the individual. Acute doses of under 100 rem may not have any noticeable immediate impact, although minor blood changes are possible. Exposures between 100 and 200 rem may induce nausea and vomiting for about one or two days. A latency period of up to two weeks follows during which the exposed individual will feel normal. Then new symptoms—mild malaise, loss of appetite, and changes in blood character—may appear. Recovery is likely in about three months unless complicated by previous poor health or other injuries. At higher doses, more life-threatening effects can be expected. Nausea, vomiting, diarrhea, loss of appetite, and malaise, all common in radiation injury, will occur. A latency period of a few days is followed by return of these symptoms, which may be accompanied by infection resulting in a steplike rise in temperature, small hemorrhages under the skin, and spontaneous bleeding in the mouth, the intestinal tract, and the kidney.[3] Assuming little or no medical treatment, approximately 10 percent of those exposed can be expected to die from midline doses of 200 to 250 rem, 50 percent from 300 to 350 rem, and 100 percent from 450 to 500 rem. The administration of supportive medical treatment, including barrier nursing, copious antibiotics, and transfusions of whole blood, packed cells, or platelets, can significantly lower the mortality threshold. See tables 1-2 and 1-3.

The effects of less acute and chronic low-level radiation are more uncertain. Some radiation exposure is inevitable from natural sources, such as

Table 1-1
Half-Lives of Radionuclides in Body Organs

Radionuclide	Radiation	Critical Organ	Half-Life Physical	Half-Life Biological	Half-Life Effective
Iodine-131	Beta	Thyroid	8 days	138 days	7.6 days
Strontium-90	Beta	Bone	28 years	50 years	18 years
Cesium-137	Gamma	Whole body	30 years	70 days	70 days
Plutonium-239	Alpha	Bone	24,400 years	200 years	198 years
		Lung	24,400 years.	500 days	500 days

Source: U.S. Atomic Energy Commission, *The Safety of Nuclear Power Reactors and Related Facilities*, WASH 1250 (Washington, D.C.: U.S. Atomic Energy Commission, July 1973), p. 4-23.

Table 1-2
Expected Effects of Acute Whole-Body Radiation Doses

Acute Exposure [within 24 hours] Roentgens (R)[a]	Probable Effect
0-50	No obvious effect, except possibly minor blood changes
80-120	Vomiting and nausea for about 1 day in 5 to 10 percent of exposed population; fatigue but no serious disability
130-170	Vomiting and nausea for about 1 day, followed by other symptoms of radiation sickness in about 25 percent of those exposed; no deaths anticipated
180-220	Vomiting and nausea for about 1 day, followed by other symptoms of radiation sickness in about 50 percent of exposed population; no deaths anticipated
270-330	Vomiting and nausea in nearly all exposed population on first day, followed by other symptoms of radiation sickness; about 20 percent deaths within 2 to 6 weeks after exposure; survivors convalescent for about 3 months
400-500	Vomiting and nausea in all those exposed on first day, followed by other symptoms of radiation sickness; about 50 percent deaths within 1 month; survivors convalescent for about 6 months
550-750	Vomiting and nausea in all those exposed within 4 hours, followed by other symptoms of radiation sickness; up to 100 percent deaths; few survivors convalescent for about 6 months
1,000	Vomiting and nausea in all those exposed within 1 to 2 hours; probably no survivors from radiation sickness
5,000	Incapacitation almost immediately; all those exposed will be fatalities within 1 week

Source: Samuel Glasstone, ed., *The Effects of Nuclear Weapons* (Washington, D.C.: U.S. Government Printing Office, 1957), p. 471.
[a]These figures approximate biological consequences in rem.

cosmic rays, uranium and thorium in the earth, and certain radioactive substances in the body. This background radiation will vary at different points on the globe, but it is usually on the order of 100 millirem per year. Although this radiation is not necessarily harmless, radiobiologists assume that it is something the body has learned to tolerate and is used as the baseline for setting standards for man-made radiation.

There is considerable controversy among scientists over the level at which man-induced radiation results in significant biological effects. According to the *Reactor Safety Study*, "Exposure to even low levels of radiation, in addition to the natural background of radiation that exists, is generally believed to increase the likelihood of certain diseases and to increase certain genetic effects."[4] Nonetheless some additional radiation is inevitable from nuclear energy. Various standards have been suggested or set as a practical threshold. For the general public, the U.S. government's Environmental Protection Agency standard is a whole body dose of 0.025 rem

Table 1-3
Probability of Death within Sixty Days

Dose Range (rem) [within 24 hours]	Minimal Treatment ("A)[a]	Supportive Teatment ("B)[b]
0-50	0	0
50-100	0	0
100-150	.0001	0
150-200	.0065	0
200-250	.11	0
250-300	.26	0
300-350	.54	.0008
350-400	.78	.02
400-450	.93	.16
450-500	.985	.38
500-550	1.0	.7
550-600	1.0	.85
600-650	1.0	.97
Over 650	1.0	1

Source: Jan Beyea, *A Study of Some of the Consequences of Hypothetical Reactor Accidents at Barsebäck*, DS I 1978:5 (Stockholm: Swedish Energy Commission), and PU/CES 61 (Princeton, N.J.: Center for Environmental Studies, Princeton University, 1978), p. II-22. Reprinted with permission.
Note: The table represents a 25 rem shift downwards of the *Reactor Safety Study* curve. Thus the curve is more conservative.
[a]Minimal treatment assumes little or no medical treatment.
[b]Supportive medical treatment includes barrier nursing, copious antibiotics, and transfusions of whole blood, packed cells, or platelets.

per year.[5] Using a more liberal standard, the International Commission on Radiological Protection (ICRP), a worldwide body of leading radiobiologists, suggests 0.17 rem per year. Allowing for greater occupational risk, its standard for radiation workers is 5 rem per year.[6]

To arrive at standards, institutions have relied on studies of the Japanese atom bomb survivors, the Marshall Islanders irradiated by bomb-test fallout in 1954, uranium miners exposed to radon, doctors and patients involved in radiotherapy, and animal experiments. Although statistical relations between doses and cancers have been computed, considerable uncertainty about the relationship remains. Thus far, the data base is limited to the first twenty to forty years after irradiation rather than a lifetime. Methodologically there are questions whether the excess risk of cancer for a given dose is a constant number of additional cases or a percentage increase; the latter would suggest higher cumulative excess cancers (that is, radio-induced cancers). Yet to be determined is whether radiation accumulated over time in small doses has the same effect as a single acute dose. The calculation is further complicated by different estimates of the effect of low-LET radiation.

The sex and age of those exposed have a complex relation to susceptibility. Women are more vulnerable to breast and thyroid cancers caused by irradiation and men to certain types of leukemias. People exposed in their youth are more likely to manifest cancers and other effects than are older people. For example, radiation-induced breast cancer is more apt to occur from exposures in adolescence and early adulthood. People under ten who absorb radioiodine are more likely to develop thyroid nodules and cancers than those over twenty. Prenatal irradiation will increase congenital malformities, notably small head size and mental retardation. Age also has an effect on the rate at which cancers manifest themselves. These maladies have a latency period that varies from years to decades and tends to be greater for older people than for younger people. With leukemia, for example, the period increases from nine to fifteen years with increasing age.[7]

Despite the uncertainties and numerous variables, estimates of the effects of radiation on people have been computed. According to one estimate a one-time, 1-rem dose to 1 million people will result in 90 to 470 excess cancer deaths. The range reflects different assumptions about relative and absolute risks and whether their duration extends for thirty years or over a lifetime. One hundred eighty deaths per million man rem often is used as an average figure, compared with a normal cancer mortality rate of 200,000 persons per million in the United States.[8] (See Appendix for some man-rem dose coefficients for delayed effects.)

At first glance, the normal cancer mortality rate might suggest that the problem posed by radiation is relatively small. However its dimensions can be appreciated only in terms of the numbers exposed and, assuming a simple consequence model, the intensity of irradiation. For illustrative purposes only, if 1 million people were subjected to 150 rem—the point at which early cancers begin to manifest—one can expect either 13,500, 27,000, or 70,500 excess late cancer deaths depending on whether a low, average, or high estimate is assumed. This represents approximately 6 percent, 13 percent, and 35 percent, respective increases in cancers. In this light, the problem is serious by any reasonable standard. Whether such a scenario is accurate is another question; this point will be addressed later.

In addition to cancers, the most prominent somatic effects are thyroid nodules, which are abnormal benign or malignant growths induced by inhalation or ingestion of radio iodine through foodstuffs. Because thyroid cancers are well differentiated, relatively slow growing, and operable, the mortality rate is much lower than for other cancers. Other effects include hypothyroidism and growth retardation.[9] Susceptibility was underscored by the manifestation of thyroid neoplasms beginning in 1963 in almost one-third of the Marshall Island inhabitants of Rongelap (twenty-seven of eighty-six) whose thyroids were exposed to 175 rd from the Bikini Island weapons test fallout in 1954. This figure included three carcinomas. Children under ten were found to be particularly vulnerable because of smaller thyroid size and rapid growth of the thyroid during these preadolescent years.[10]

Genetic effects are more difficult to measure than are somatic ones because they may not manifest for generations and because radio-induced effects are difficult to distinguish from the estimated 20 percent contribution that genetics already makes to ill health. To date no such effects have been measured. A recent study of the children of Japanese survivors of the atom bomb uncovered no significant effect except for those exposed in utero. A report by the National Academy of Sciences found that there is "no direct evidence of human genetic effects, even at high doses." "Nevertheless," it believes,

> the animal evidence is so overwhelming that we have no doubt that humans are affected in much the same way. In contrast to somatic effects, where the concern is concentrated mainly on malignant disease, the genetic effects are on all kinds of conditions—for the spectrum of radiation-caused genetic disease is almost as wide as the spectrum for all other causes.[11]

Among the potential consequences are gene mutations distinguished as dominant, recessive, and sex-linked types. Some may be trivial and invisible, others conspicuous and lethal. Dominant mutations may occur in the first generation. Some 415 such abnormalities exist, with some 528 others less well established, including polydactyly (extra fingers and toes), achondroplasia (short-limbed dwarfism), Huntington's chorea (progressive involuntary movements and mental deterioration), one type of muscular dystrophy, and several types of anemia. Recessive mutations may take many generations to manifest. There may be as many as 783 such diseases, including Tay-Sach's disease (blindness and death in the first few years of life), sickle cell anemia, and cystic fibrosis. Sex-linked disorders may appear relatively early as recessive mutations. They include color-blindness and hemophilia. In addition, genetic mutations include chromosomal aberrations or broken chromosomes, resulting in mongolism, embryonic death, physical abnormalities, and mental deficiency. Finally, genetics contributes to diseases of multifactoral origins, including heart disorders, epilepsy, schizophrenia, asthma, and diabetes.[12]

As in the case of somatic effects, several genetic estimates per radiation dose have been calculated (table 1-4). The most recent Ford Foundation study suggests that 214 to 5,200 disease, that is, man-induced cases can be expected for a 5 rem exposure per generation per million live births. This estimate compares with a current incidence of over 94,000 per million live births.

Table 1-4
Estimates of First-Generation Genetic Effects

Type of Disorder	BEIR		WASH-1400		Possible Revisions[a]	
	Current Incidence	Disease Cases	Current Incidence	Disease Cases	Current Incidence	Disease Cases
Single gene disorders						
Autosomal dominance	10,000	50-500	10,000	100	3,000	60-600
X-chromosome link	400	0-15	400	3	400	7-70
Autosomal recessives	1,500	0	1,500	0	1,500	0
Multifactorial disorders	40,000	5-500	40,000	10-100	85,000	42-4,200
Effects of chromosome abberations						
Unbalanced rearrangements	1,000	60	1,000	60	500	100-200
Aneuploidy	4,000	5	4,000	5	4,000	5-130
Total		120-1,080		178-268		214-5,200
For 3 × 10^6 live births (U.S. population)		360-3,240		534-804		642-15,600
Risk estimates (first generation defects per man-rein)		12-100 × 10^{-6}		18-27 × 10^{-6}		21-5,200 × 10^{-6}

Source: From *Nuclear Power Issues and Choices*, Copyright 1977, The Ford Foundation. Reprinted with permission from Ballinger Publishing Company.

Note: Based on 5 rem per generation per million live births.

[a]Revisions suggested by the Nuclear Energy Policy Study Group.

In addition it is reasonable to expect psychological and sociological consequences from radiation exposure. Emotional distress has afflicted many people who live near the Three Mile Island reactor in Pennsylvania. Whether this will linger is uncertain. The traumatic experience of Japanese bomb survivors suggests cause for concern, at least when contamination is significant. Although irradiation resulting from nuclear facility destruction will not be as dramatic—there will be no comparable fireball and shock wave—the increasing psychological sensitivity worldwide to nuclear facility accidents may result in effects similar to those suffered by the Japanese.[13] Robert Jay Lifton documented these in *Death in Life*.

On the basis of interviews, Lifton found that many, if not most, survivors continued to be traumatized by their experience almost two decades after the event. They showed a preoccupation with death and dying and a sense of guilt at having survived. Anxieties about the connection of ill health and irradiation were common even among people who at the time of exposure were too young to remember the explosion and aftermath and among children of survivors born in later years. The anxiety was reinforced by the varying latent periods of different cancers. When the incidence of leukemia peaked in the period 1950-1952, there was a feeling that the worst might be over. However, when other cancers subsequently appeared, survivors felt that they faced a never-ending threat to life and wellbeing, raising concerns about sexual adequacy and the danger to their children. The comments of two interviewees are instructive in human terms. A young company executive said:

> Even when I have an illness which is not at all serious—as for instance, when I had very mild liver trouble—I have fears about its cause. Of course, if it is just an ordinary condition, there is nothing to worry about, but if it has a direct connection to radioactivity, then I might not be able to expect to recover. At such times I feel myself very delicate. . . . This happened two or three years ago, I was working very hard and drinking a great deal of sake at night in connection with business appointments, and I also had to make many strenuous trips. So my condition might have been partly related to my using up so much energy in all of these things. . . . The whole thing is not fully clear to me . . . but the results of statistical study show that those who were exposed to the bomb are more likely to have illnesses—not only of the liver, but various kinds of new growths, such as cancer or blood diseases. My blood was examined several times but no special changes were discovered. . . . When my marriage arrangements were made, we discussed all these things in a direct fashion. Everyone knows that there are some effects, but in my case it was the eleventh year after the bomb, and I discussed my physical condition during all of that time. From that, and also from the fact that I was exposed to the bomb while inside of a building and taken immediately to the suburbs, and then remained quite a while outside the city—judging from all of these facts, it was concluded that there was very little to fear concerning my condition. . . . But in general there is a great concern that people who were

exposed to the bomb might become ill five or ten years later or at any time
in the future. . . . Also when my children were born, I found myself worry-
ing about things that ordinary people don't worry about, such as the
possibility that they might inherit some terrible disease from me. . . . I
heard that the likelihood of our giving birth to deformed children is greater
than in the case of ordinary people . . . and at that time my white blood
cell count was rather low. . . . I felt fatigue in the summertime and had a
blood count done three or four times. . . . I was afraid it could be related
to the bomb, and was greatly worried. . . . Then after the child was born,
even though he wasn't a deformed child, I still worried that something
might happen to him afterward. . . . With the second child, too, I was not
entirely free of such worries. . . . I am still not sure what might happen,
and I worry that the effects of radioactivity might be lingering in some way.[14]

Commenting on this case, Lifton concluded that although the man carries
out his life effectively and has essentially good health and normal children,
he is plagued by anxieties about his health, his marriage, and the effect of
his irradiation on his children. "Each hurdle is surmounted, only to reap-
pear in new form."[15] A grocer expressed these feelings in still stronger
terms:

Frankly speaking, even now I have fear. . . . Even today people die in the
hospitals from A-bomb disease, and I worry that I too might sooner or
later have the same thing happen to me. . . . So when I hear about people
who die from A-bomb disease, or who have operations because of this illness, then I feel that I am the same kind of person as they.[16]

The psychological trauma is further reinforced by social ostracism.
Many survivors, tainted by death, feel like outcasts and are looked upon
differently. They have difficulty finding jobs because employers are
suspicious of their health. Consequently survivors find themselves at a
lower socioeconomic level of society, feeling that in work as in other aspects
of social affairs they are impaired.[17]

As long as nuclear materials are isolated, they pose no health hazard.
However, once released there are three principal and two lesser routes to
human contamination. The first is external radiation from a damaged
nuclear facility's released products. The *Reactor Safety Study* suggests that
such exposure from power plants will occur over a period of thirty minutes
to a few hours. Internal radiation from inhaled radionuclides takes place
over the same time frame. The doses accumulated depend upon the physical
and biological decay and removal processes defined in terms of the effective
half-life. Inhalation is the most important contributor to thyroid and lung
disease. External irradiation from material deposited on the ground (the
ground dose) is the third, and usually most important, contributor to early
fatalities and long-term health effects.[18] Figure 1-2 shows a view of each of
these routes.

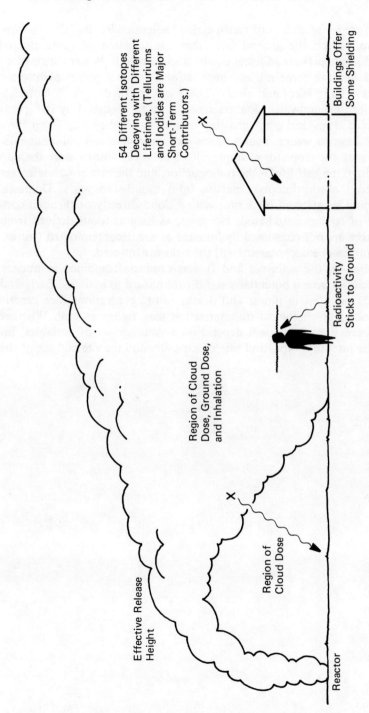

54 Different Isotopes Decaying with Different Lifetimes. (Telluriums and Iodides are Major Short-Term Contributors.)

Buildings Offer Some Shielding

Radioactivity Sticks to Ground

Region of Cloud Dose, Ground Dose, and Inhalation

Region of Cloud Dose

Effective Release Height

Reactor

Figure 1-2. Side View of Radioactive Plume

Source: Jan Beyea, "The Impact in New York City of Reactor Accidents at Indian Point," Statement to the New York City Council, June 11, 1979, corrected June 20, 1979 (Princeton, N.J.: Center for Energy and Environmental Studies, 1979), p. 29. Reprinted with permission.

Exposure to inhalation of resuspended radionuclides that already have been deposited on the ground and from consumption of contaminated water and foodstuffs are of lesser significance to health. Water may be contaminated because some nuclear installations, notably power plants, are located near large lakes and rivers. There is also the danger of materials reaching into groundwater. Radioaerosols may be absorbed by the leaves and stems of crops during the growing season, where they may linger for a period of days to weeks; root systems can assimilate soil contaminants. Depending on the crop (for example, rice and wheat absorb more than do other grains), the half-life of the radionuclide, and the rate at which it fixes in soil, root absorption may persist from months to years. Domestic animals may be affected by a radioactive cloud directly or through consumption of contaminated feed. However, as long as foods derived from such sources are not consumed by humans or are decontaminated (for example, milk can be decontaminated) the risks are lowered.[19]

To sum up, the actinides and fission products contained in nuclear energy facilities pose a potentially significant hazard to humans. Early and late effects manifest in illness and death; future generations face genetic consequences; psychological traumatization may follow as well. Whether these repercussions occur will depend on a complex set of variables, including the intensity, type, and rate of exposure and the age and sex of the exposed.

2

The Vulnerability of Nuclear Facilities to Acts of War

The Nuclear Reactor

Atomic power plants are the most conspicuous portion of the nuclear fuel cycle. By late 1979, 527 were in operation, under construction, or on order in thirty-six countries (table 2-1). Table 2-2 suggests that by 1985 nuclear energy will represent at least 15 percent of the electrical capacity of nine countries, whereas only one nation had achieved this capacity in 1978. But these estimates may be too large. For example, in 1978 the Organization for Economic Cooperation and Development (OECD) projections for Austria, Belgium, Canada, Denmark, Finland, France, the Federal Republic of Germany, Italy, Japan, Luxembourg, the Netherlands, Portugal, Spain, Sweden, Switzerland, the United Kingdom, and the United States were revised considerably downward (table 2-3). Still substantial increases for these countries are foreseen, and this is likely to apply to other nations as well.

Incentives to Destroy Atomic Power Plants in War

The incentives to destroy nuclear facilities in war particularly apply to reactors. The atomic reactor is central to the nuclear fuel cycle; it is the locus of energy production. Without energy, machines cannot work to support the military. The logic of this argument is obvious, though uncommonly applied, often resulting in needlessly prolonged conflict. During World War II, power plants were only secondary targets in both German and Allied attacks. Their dispersal, the inaccuracy of ammunitions, and the high cost of the large number of bombs necessary to destroy these targets explain this policy partially. However, it appears that poor planning may be the principal reason for it. After the war, the United States assessed the effectiveness of its military strategy on the basis of field observations, German documents, and interviews with German industrialists and technocrats. A chief electrical engineering designer summed up the German view of the Allied oversight:

> The war would have finished two years sooner if you had concentrated on the bombing of our power plants earlier. The best plants to bomb would

Table 2-1
Nuclear Power Plants 30MW(e) and Over, December 31, 1979

Country	Operating[a]	Under Construction[b]	Ordered[c]	Total
Argentina	1	1	0	2
Austria	0	1[d]	0	1
Belgium	3	4	0	7
Brazil	0	2	1	3
Bulgaria	2	2	0	4
Canada	10	9	4	23
Czechoslovakia	2	4	3	9
Egypt	0	0	1	1
Finland	2	2	0	4
France	15	28	8	51
German Democratic Republic	4	3	0	7
German Federal Republic	12	9	7	28
Hungary	0	4	0	4
India	3	5	0	8
Iran	0	2	0	2
Iraq	0	0	1	1
Italy	4	3	2	9
Japan	22	7	1	30
Libya	0	0	1	1
Luxembourg	0	0	1	1
Mexico	0	2	0	2
Netherlands	2	0	0	2
Pakistan	1	0	0	1
Philippines	0	1	1	2
Poland	0	0	2	2
Roumania	0	0	2	2
South Africa	0	2	0	2
South Korea	1	6		7
Spain	3	8	4	15
Sweden	6	5	1	12
Switzerland	4	1	2	7
Taiwan	2	4	0	6
Turkey	0	0	1	1
United Kingdom	33	6	0	39
United States	70	88[e]	31	189
U.S.S.R.	25	16	[f]	41
Yugoslavia	0	1	0	1
Total	227	226	74	527

Source: "World List of Nuclear Power Plants," *Nuclear News* 22, 2 (February 1980):67-86.

[a]Units in commercial operation.

[b]Includes plants that may be constructed but are not yet in operation.

[c]The criterion for listing a unit is either an order or letter of intent signed for the reactor. In cases where the definition of "letter of intent" is ambiguous or where a special situation exists, inclusion depends on judgment of the utility.

[d]A plebiscite has deferred Austria's nuclear energy program indefinitely.

[e]Includes the damaged Three Mile Island reactor.

[f]Insufficient information.

have been the steam plants. Our own air force made the same mistake in England. They did not go after English power plants and they did not persist when they accidentally damaged a plant. Your attacks on our power plants came too late. This job should have been done in 1942. Without our public utility power plants we could not have run our factories and produced war materials. You would have won the war and would not have had to destroy our towns. Therefore, we would now be in a much better condition to support ourselves. I know the next time you will do better.[1]

Such opinions were typical, contributing to the conclusion of U.S. analysts that

had electric utility plants and large substations and the larger industrial power plants been made primary targets as soon as they could have been brought within the range of Allied strategic bombing attacks, all evidence indicates that the destruction of such installations would have had a catastrophic effect on Germany's war production.[2]

The Germans similarly erred in their campaign against the Soviet Union. When war on the eastern front began, the Luftwaffe was instructed not to attack Soviet industry, including power plants, so the facilities could be exploited under German control. Instead it concentrated on air superiority and ground support missions. As its fortunes turned, Germany reconsidered destruction of Soviet industry and gave top priority to a large tank factory and the Soviet rail network. By the time the power industry was considered for destruction, German retreats made the long-range bombing impossible. Commando attacks and one-way aircraft suicide missions were never carried out.[3]

More recently power plants have been prime targets in wartime. In Korea, these attacks came after two years of conflict. The delay was due to an early decision by the United States not to destroy large Yalu River hydroelectric dams serving both China and North Korea in order to avoid giving Peking an excuse to intervene. The decision was reversed in June 1952 when negotiations deadlocked and destruction of the plants seemed necessary to hasten the war's conclusion and to make more difficult the repair work the communists were doing in small industrial establishments and railway tunnels.

In the Middle East during the 1973 war, Israeli planes destroyed power stations at Damascus and Homs, Syria, to subdue Syrian military activity and to deter other countries from entering the conflict. In Vietnam, the United States destroyed some electrical facilities, but these were not primary targets given their small size and Hanoi's limited industry and reliance on imports from abroad. In none of these cases have the implications of energy destruction been subject to the same scrutiny as with World War II.[4]

Table 2-2
Nuclear Generating Capacity Outside the United States

Country	1978			1979		1985		1990		2000	
	MW(e) Installed	%of Capacity	% of Generation	MW(e) Installed	% of Capacity	MW(e) Installed	% of Capacity	MW(e) Installed	% of Capacity	MW(e) Installed	% of Capacity
Argentina	344	3.0	8.0	344	3.0	944	8.5	1,642	12.0	n.a.	23.0
Austria	0	0	0	0	0	n.a.	n.a.	n.a.	n.a.	n.a.	n.a.
Belgium	1,667	16.0	24.5	1,667	16.0	5,427	38.0	n.a.	n.a.	n.a.	n.a.
Brazil	0	0	0	0	0	626	n.a.	10,586	13.0	75,000	40.0
Bulgaria	880	n.a.	n.a.	880	n.a.	1,760	n.a.	5,760	35.0	n.a.	n.a.
Canada	4,774	8.0	8.8	5,514	8.0	9,700	10.0	16,900	15.0	45,200	30.0
Chile	0	0	0	0	0	0	0	600	15.0	n.a.	n.a.
China, People's Republic of	0	0	0	0	0	0	0	n.a.	n.a.	n.a.	n.a.
China, Republic of (Taiwan)	636	9.0	7.3	1,212	16.6	4,928	31.0	n.a.	n.a.	n.a.	n.a.
Cuba	0	0	0	0	0	880	n.a.	n.a.	n.a.	n.a.	n.a.
Czechoslovakia	112	n.a.	n.a.	112	n.a.	4,952	17.0	0	32.1	n.a.	50.0
Denmark	0	0	0	0	0	0	0	0	n.a.	n.a.	n.a.
Egypt, Arab Republic of	0	n.a.	0	0	0	0	0	600	n.a.	6,900	28.6
Finland	420	n.a.	n.a.	1,080	10.0	2,160	20.0	5,000	6.7	10,000	40.0
France	8,330	10.0	12.5	8,330	10.0	33,125	50.0	n.a.	30.0	n.a.	85.0
Germany, Democratic Republic of	1,400	n.a.	5.0	1,400	n.a.	5,360	n.a.	26,580	65.0	12,000	50.0
Germany, Federal Republic of	5,893	10.0	12.6	8,887	12.0	19,534	n.a.	4,760	n.a.	6,000	n.a.
Hungary	0	0	0	0	0	1,760	n.a.	2,116	10.0	n.a.	48.0
India	596	2.5	2.3	596	2.5	1,676	4.0	2,900	4.5	n.a.	6.0
Indonesia	0	0	0	0	0	1,600	n.a.	0	n.a.	3,800	n.a.
Israel	0	0	0	0	0	0	0	0	n.a.	n.a.	n.a.
Italy[a]	1,412	3.3	2.5	1,412	3.2	2,434	4.5	13,400	0	78,000	30.0
Japan[b]	11,502	9.9	10.1	14,952	12.2	30,000	16.7	53,000	14.8	46,416	n.a.
Korea, Republic of South	587	8.5	11.5	587	8.5	3,815	20.0	n.a.	n.a.	n.a.	n.a.
Libya	0	0	0	0	0	0	0	n.a.	n.a.	n.a.	n.a.
Luxembourg	0	0	0	0	0	1,300	n.a.	n.a.	n.a.	n.a.	n.a.
Mexico	0	0	0	0	0	1,308	5.0	n.a.	n.a.	n.an	n.a.
The Netherlands	505	3.3	7.0	505	3.3	505	2.7	1,505	7.3	3,505	17.7
Pakistan	125	n.a.	n.a.	125	n.a.	725	n.a.	n.a.	n.a.	n.a.	n.a.

The Philippines, Republic of	0	0	0	0	620	10.5	620	7.2	1,240	6.7
Poland	0	0	0	0	440	n.a.	4,880	9.0	n.a.	n.a.
Portugal	0	0	0	0	0	0	0	0	5,000	30.0
Romania	0	0	0	0	440	n.a.	n.a.	20.0	n.a.	n.a.
South Africa, Republic of	1,082	8.0	1,082	3.9	1,844	7.5	3,000	9.0	7,000	12.0
Spain[c]	3,700	25.0	3,700	14.6	n.a.	n.a.	n.a.	n.a.	n.a.	n.a.
Sweden	1,006	18.8	1,926	17.5	8,380	28.0	9,430	n.a.	n.a.	n.a.
Switzerland	0	0	0	0	2,871	21.9	900	10.0	n.a.	n.a.
Thailand					0	0	660	4.0	n.a.	n.a.
Turkey							n.a.	25.0	n.a.	n.a.
Union of Soviet Socialist Republics	7,905	2.0	9,905	4.3	34,135	10.0	12,836	13.0	[d]	33.0[d]
United Kingdom[d]	6,426	13.0	6,426	10.0	10,196	11.0	n.a.	n.a.	n.a.	n.a.
Yugoslavia	0	0	0		632	n.a.				n.a.

Source: Atomic Industrial Forum, *INFO: News Release*, February 6, 1980. Reprinted with permission.

[a]Percentage of capacity for 1985 and 1990 includes share of Super Phenix.

[b]Gross MW(e) used. 1978 figures for fiscal year April 1, 1978-March 3, 1979.

[c]Estimate for 1987: 10,500 MW(e), or 22.0 percent.

[d]1978 figures for fiscal year April 5, 1978-April 4, 1979. Estimates for 2000: 27,400-40,200 MW(e), or 20.9-37.9 percent.

Table 2-3
Nuclear Capacity Growth Estimates in OECD Countries
(in GW(e)ª)

Forecast as of	1985		1990		2000	
	High	Low	High	Low	High	Low
End of 1977	343	259	640	459	1,640	850
End of 1978	280	238	483	388	1,021	681

Source: Organization for Economic Cooperation and Development, Nuclear Energy Agency, *Seventh Activity Report: 1978* (Paris: Organization for Economic Cooperation and Development, 1979), p. 16.
ªGW(e): one thousand million watts, 1000 MW(e).

Disruption of energy production does not necessarily require destruction of the boiler (or in the case of the atomic power plant, the nuclear reactor). In the energy-producing cycle there are a number of vulnerable components whose destruction could disrupt production for weeks or even months: the main turbine generator, the electric generator, conductors through which power flows to transformers, the transformers themselves, which step up plant voltage for transmission, and high-tension circuit breakers and switches through which the power is taken to outgoing lines.[5] Disruption of these components is sufficient to stop production without releasing radionuclides, but accurate bombardment and good command and control are vital and enemy satisfaction with this limited destruction is assumed.

Destruction of power plants for other reasons than eliminating energy production is more likely to release contaminants. Theodore Taylor, a noted physicist, points out the possibility that a nation may regard the civil nuclear energy plant of an adversary as a guise for a nuclear weapons program and thus try to sabotage or destroy it at the outbreak of war. South Africa's current nuclear energy program, to which some ascribe military motivations, and oil-rich Libya's desire to acquire a power plant may be cases in point. Taylor also suggests that a nation or possibly a subnational group might destroy a plant in order to escalate a conflict between two other parties. He points to the possibility that tensions between China and the Soviet Union could develop into a border war but one in which both sides restrain their use of nuclear weapons.[6] It might be tempting for a third entity—perhaps Taiwan—to sabotage a power plant to increase hostilities. Taylor's scenario could be applied to other areas as well.

In each of these cases, the release of radionuclides may be an incidental by-product of destruction, but it may also be the primary objective. Precedent for mass destruction does exist. Fire, herbicides, and flood have made effective weapons of mass destruction. Samson released several hundred foxes with their tails burning to set afire the Philistines' agriculture and horticultural fields. American incendiary bombing of Japan and Germany during World War II was the most costly use of fire in terms of human lives lost. The Boers used incendiaries to destroy crops during the 1899-1902 war with England; and the British did likewise in their counterinsurgency campaign in Malaya in the 1950s. In Vietnam, the United States used incendiaries as well as herbicides and tractors with large blades, called Rome ploughs, to destroy crops and forest cover.

Flooding caused by destruction of existing levees, dikes, or dams has at times been even more devastating than fire. In its conflict with Japan during the late 1930s, China destroyed a dike on the Yellow River, drowning several thousand Japanese soldiers and stopping their advance along the front, but it cost the Chinese several million inundated hectares of farmland and destruction of thousands of villages, resulting in the displacement of millions of Chinese. During World War II the Dutch destroyed their dikes to hamper the German invasion. As the war turned against Germany, the Germans flooded large tracts of Dutch land as they withdrew from the country. In the Korean war, the United States attacked irrigation dams in the north, causing widespread devastation. It also bombed dams, dikes, and seawalls during the Vietnam war, albeit inadvertantly according to official accounts. In addition, it employed weather modification techniques to increase rainfall during the wet season, making military operations for its adversary more difficult.[7]

Nuclear power plants also have enormous inherent value. Taylor noted that

> a large nuclear power plant represents one of the greatest concentrations of high value in a small volume that exists in the modern world. The capital cost of a 1000 Mw(e) nuclear power plant is over $100 million. [Today, over ten years after Taylor's writing, some 1000 Mw(e) plants cost over $1 billion.] It is difficult to think of other targets of comparable value that might be rendered permanently useless by the explosion of a few pounds of high explosives in the right place.[8]

Destruction would significantly delay the postwar recovery of any state that heavily relies on nuclear energy.

The Vulnerability of Nuclear Power Plants

To comprehend the vulnerability of reactors to acts of war requires some familiarity with their operation.[9] In general, tons of processed uranium, thorium, and/or plutonium are fabricated into pellets, spheres, or balls and placed in casings called cladding and assembled to form the core. Interspersed among the assemblies are neutron-absorbing control rods. Fission—the splitting of the atomic nucleus that results in a release of energy—begins with their removal. To help sustain the process, moderators in the form of water, deuterium oxide (heavy water), or graphite are dispersed within the core of thermal reactors. They are not present in more compact fast or breeder reactors. Resulting energy manifests as heat, which is transported by air, carbon dioxide, helium, water, or molten metal to water, which generates steam to drive electricity-producing turbines. These elements also cool the core and must continue to do so even when the reactor is not in operation to remove decay heat continuously generated (at a declining rate) by fission products. Although this heat will be a small fraction of what it is when the reactor is in operation, without the coolant, the core will melt, breach containments, and release radioactive products into the environment.

There are at least twelve different types of energy-producing reactors in operation around the world. They are distinguished by fuels, moderators, control systems, cooling arrangements, and configurations. They include the advanced gas-cooled reactor (AGR), boiling water reactor (BWR), gas-cooled, heavy-water-moderated reactor (GCHWR), gas-cooled reactor (GCR), high-temperature gas-cooled reactor (HTGR), heavy-water-moderated, boiling light-water-cooled reactor (HWLWR), light-water-cooled, graphite-moderated reactor (LGR), liquid metal fast-breeder reactor (LMFB); light-water-cooled heavy-water-moderated and cooled reactor (LWCHW); pressurized heavy-water-moderated and cooled reactor (PHWR); pressurized water reactor (PWR); and thorium high-temperature reactor (THTR).[10]

Roughly 80 percent of the world's reactors are moderated and cooled by ordinary light water. Therefore they will be the focal point for this review of reactor vulnerability. Figures 2-1 and 2-2 schematize the two most common configurations: the boiling water reactor and the pressurized water reactor. In both designs, thousands of tubes containing fuel pellets are bunched together in a pressure vessel.

In the BWR, water boils into steam in the vessel and is transported by pipes to turbine generators. It is then condensed with cooling water and recycled back to the reactor. In the PWR, water is circulated through the core under great pressure (2,000 lbs./sq.in.; this compares to 1,000 lbs./sq.in. for the BWR). The high pressure of the PWR allows water to heat to 600 degrees without boiling. The heat is transported through pipes to a steam generator, where it boils water circulating at a lower temperature and a pressure creating steam to drive turbine generators. In both designs, large pumps and pressurizers circulate the water.

As long as there are no major coolant pipe breaks, pressure vessel ruptures, mismatches of power and coolant due to excessive fission or undercooling, or failure to control reactivity in the raising or lowering of the plant's output, the reactor will operate without serious problem. To guard

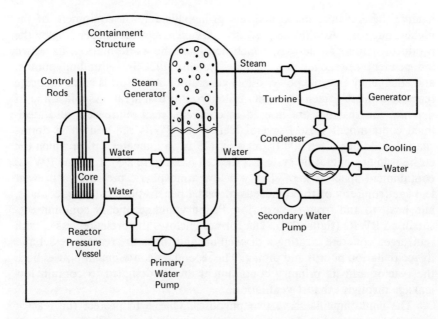

Figure 2-1. Schematic Diagram of a Pressurized-Water Reactor Power System

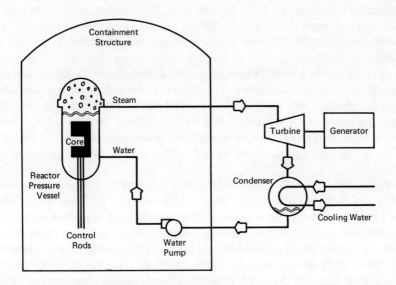

Figure 2-2. Schematic Diagram of a Boiling-Water Reactor Power System

against these events, manufacturers principally rely on equipment of the highest integrity. Additional precautionary measures vary depending on the producer. American designs, which dominate the world market, have two independent emergency core cooling systems (ECCS), redundant pumps, and emergency external power diesel generators. An effort is made to locate reactors at least fifteen to twenty miles from urban areas. As a last line of defense, the reactors are housed in concrete, steel-reinforced, and steel-lined containment vessels several feet thick. PWRs are housed in dome-shaped buildings often 200 ft. high by 125 ft. in diameter within which the entire primary operating system is located (see figure 2-3). In some PWRs, containment space is kept slighty below atmospheric pressure to prevent leakage from any of the hundreds of vessel penetrations carrying coolant, and heat, to and from the exterior. Primary and secondary containments surround BWRs (figure 2-4). The former encloses the pressure vessel, with reinforced concrete creating a drywell around the entire reactor vessel and its recirculation pumps and piping. The secondary containment houses both the reactor and its primary containment and is designed to contain low leakage through exhaust ventilation.

The containments serve two purposes. One is to protect the reactor against external events, such as extreme meteorological conditions, tank car explosions, and aircraft crashes. The second is to prevent the release of radionuclides in the event of an accident. For example, in a loss of coolant accident (LOCA), primary barriers are designed to withstand the pressure created by expelled primary coolant flashing into steam and pressurizing the containment volume. To facilitate the task, PWRs have cold water sprays and ice to condense the steam. In BWRs, the drywell channels steam into a pressure-suppression chamber where it would be condensed by a pool of water.

A spectrum of precautionary measures distinguishes American designs from those of other countries. For example, German reactors maintain greater redundancies. They are surrounded by a double primary containment, the inner being somewhat less stress resistant than the outer. Some designs have as many as four emergency core cooling systems. Plants built by other West European countries fall within the German-American spectrum.[11]

By contrast, Soviet designs place greater reliance on component dependability and less on redundancy. At the official level, the Russians believe that a nuclear power plant accident that would significantly contaminate the environment is impossible. Although some precautionary measures are taken—for example, coolant loops are generally isolated in concrete to contain leaks and plants are provided with emergency external power generators—some of the most notable safeguards that characterize European and American practices are absent. For example, the Russians do not

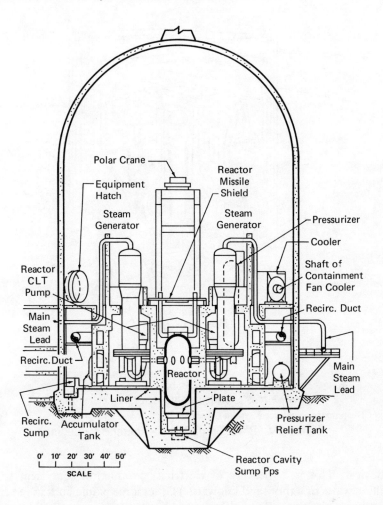

Source: American Physical Society Study Group on Light-Water Reactor Safety, *Review of Modern Physics* 47, supplement no. 1 (Summer 1975):S18. Reprinted with permission.

Note: From unit 2, Diablo Canyon Pacific Gas and Electric Company.

Figure 2-3. Typical PWR Containment

require a population exclusionary zone around plants; instead towns for workers are planned in close proximity. Reactors do not include emergency core cooling systems and pressure-suppression mechanisms. And until recently, industrial buildings rather than containment vessels enclosed reactors. However, this policy may be changing. At least one 1,000 Mw(e) facility, under construction at Novovoronizh, is being built with a containment vessel 1m thick with an 8-mm steel liner. (Mw(e) equals Megawatt

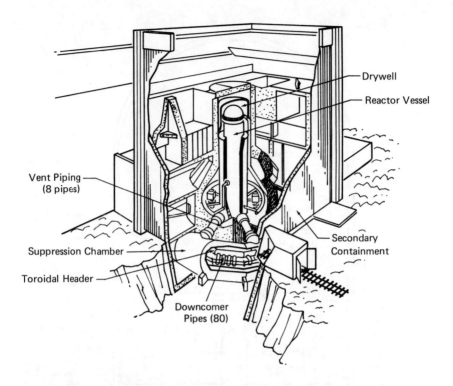

Source: *Review of Modern Physics* 47, p. S21. Reprinted with permission.
Note: The primary containment system is enclosed.

Figure 2-4. BWR Secondary Containment Building

electrical. This is a measure of electricity production equivalent to one million watts or a thousand kilowatts.) Other large plants built in the future may be so enclosed. Recent Soviet publications also suggest that at least some Russian scientists believe that future atomic facilities, including reactors as well as fuel fabrication, reprocessing, and waste treatment facilities, should be concentrated in remote nuclear parks for ecological reasons.[12]

The vulnerability of nuclear power plants can be extrapolated from the *Reactor Safety Study*'s accident sequences and a Sandia Laboratory review of their susceptibility to sabotage. Both works underscore the point that in American designs, a significant radionuclide release can occur only if both primary and compensatory systems are upset. According to the Sandia report this requires disruption of two or more of the following trains. The destruction of just one merely reduces the defense.

1. Reactor coolant makeup and decay heat removal system: auxiliary feedwater system, residual heat-removal system, emergency core cooling system, reactor coolant makeup system.

2. Steam and feedwater lines between the reactor vessel or steam generator and the check or isolation valves.
3. Reactor coolant makeup and cooling water systems, including: condensate storage tank, refueling water storage tank, suppression pool, closed cycle cooling water (component cooling) systems; open cycle cooling water (emergency service water) systems.
4. Emergency power and air supplies, including emergency DC electrical power systems, emergency AC electrical power systems, emergency air system.
5. Reactor protection system.
6. Engineered safety features in initiation system.
7. Control room.
8. Containment.
9. Containment isolation systems.
10. Containment heat-removal systems.
11. Controls instrumentation and cabling for the above systems.[13]

To this list can be added puncture of the pressure vessel.

The fact that two or more trains are required to release products underscores the importance of redundant safeguards, suggesting that some West European designs are more resistant to attack than are American designs. By contrast, Soviet designs are the most vulnerable. The Sandia findings also underline the fact that releases can result from disruptions of systems outside the containment. Of particular concern is off-site power. According to the report:

> (1) sabotage resulting in total loss of off-site power to a nuclear unit could be accomplished with relative ease according to many scenarios, (2) the sabotage could be accomplished from outside the nuclear plant security perimeter, and (3) it is impossible to prevent loss of off-site power to a nuclear unit from acts of sabotage and impractical to attempt to design to prevent them.[14]

Another problem is the disruption of primary and redundant coolant carried by pipes from the outside.

The *Reactor Safety Study* categorized the magnitude of releases largely on the failure of safeguards to perform effectively. Cited below, the scenarios suggest what willful acts of destruction must achieve to release different fractions of a facility's contents. Releases range from substantial—up to 70 percent—to minor discharges—thousandths of a percentile. Tables 2-4 through 2-7 present data on these releases in sabotage scenarios that induce a loss of coolant or some other departure from normal operation (transient) that results in a meltdown.

PWR 1

This release category can be characterized by a core meltdown followed by a steam explosion on contact of molten fuel with the residual water in the reactor vessel. The containment spray and heat-removal systems are also assumed to have failed and, therefore, the containment could be at a pressure above ambient at the time of the steam explosion. It is assumed that the steam explosion would rupture upper portion of the reactor vessel and breach the containment barrier, with the result that a substantial amount of radioactivity might be released from the containment in a puff over a period of about ten minutes. Due to the sweeping action of gases generated during the containment-vessel melt-through, the release of radioactive materials would continue at a relatively low rate thereafter. The total release would contain approximately 70% of the iodines and 40% of the alkali metals present in the core at the time of release. Because the containment would contain hot pressurized gases at the time of failure, a relatively high release rate of sensible energy from the containment could be associated with this category. This category also includes certain potential accident sequences that would involve the occurrence of core melting and a steam explosion after containment rupture due to overpressure. In these sequences, the rate of energy release would be lower, although still high.

PWR 2

This category is associated with the failure of core cooling systems and core melting concurrent with the failure of containment spray and heat-removal systems. Failure of the containment barrier would occur through overpressure, causing a substantial fraction of the containment atmosphere to be released in a puff over a period of about 30 minutes. Due to the sweeping action of gases generated during containment vessel melt-through, the release of radioactive material would continue at a relatively low rate thereafter. The total release would contain approximately 70% of the iodines and 50% of the alkali metals present in the core at the time of release. As in PWR release category 1, the high temperature and pressure within containment at the time of containment failure would result in a relatively high release rate of sensible energy from the containment.

PWR 3

This category involves an overpressure failure of the containment due to failure of containment heat removal. Containment failure would occur prior to the commencement of core melting. Core melting then would cause radioactive materials to be released through a ruptured containment barrier. Approximately 20% of the iodines and 20% of the alkali metals present in the core at the time of release would be released to the atmosphere. Most of the release would occur over a period of about 1.5 hours. The release of radioactive material from containment would be caused by the sweeping action of gases generated by the reaction of the molten fuel with concrete. Since these gases would be initially heated by contact with the metal, the rate of sensible energy release to the atmosphere would be moderately high.

PWR 4

This category involves failure of the core cooling system and the containment spray injection system after a loss-of-coolant accident, together with a concurrent failure of the containment system to properly isolate. This would result in the release of 9% of the iodines and 4% of the alkali metals present in the core at the time of release. Most of the release would occur continuously over a period of 2 to 3 hours. Because the containment recirculation spray and heat-removal systems would operate to remove heat from the containment atmosphere during core melting, a relatively low rate of release of sensible energy would be associated with this category.

PWR 5

This category involves failure of the core cooling systems and is similar to PWR release category 4, except that the containment spray injection system would operate to further reduce the quantity of airborne radioactive material and to initially suppress containment temperature and pressure. The containment barrier would have a large leakage rate due to a concurrent failure of the containment system to properly isolate, and most of the radioactive material would be released continuously over a period of several hours. Approximately 3% of the iodines and 0.9% of the alkali metals present in the core would be released. Because of the operation of the containment heat-removal systems, the energy release would be low.

PWR 6

This category involves a core meltdown due to failure in the core cooling systems. The containment sprays would not operate, but the containment barrier would retain its integrity until the molten core proceeded to melt through the concrete containment base mat. The radioactive materials would be released into the ground, with some leakage to the atmosphere occurring upward through the ground. Direct leakage to the atmosphere would also occur at a low rate prior to containment-vessel melt-through. Most of the release would occur continuously over a period of about ten hours. The release would include approximately 0.08% of the iodines and alkali metals present in the core at the time of release. Because leakage from containment to the atmosphere would be low and gases escaping through the ground would be cooled by contact with the soil, the energy release rate would be very low.

PWR 7

This category is similar to PWR release category 6, except that containment sprays would operate to reduce the containment temperature and pressure, as well as the amount of airborne radioactivity. The release would involve 0.002% of the iodines and 0.001% of the alkali metals present in the core at the time of release. Most of the release would occur over a period of 10 hours. As in PWR category 6, the energy release would be very low.

PWR 8

This category approximates a PWR design basis accident (large pipe break), except that the containment would fail to isolate properly on demand. The other engineered safeguards are assumed to function properly. The core would not melt. The release would involve approximately 0.01% of the iodines and 0.05% of the alkali metals. Most of the release would occur in the 0.5-hour period during which containment pressure would be above ambient. Because containment sprays would operate and core melting would not occur, the energy release rate would also be low.

PWR 9

This category approximates a PWR design basis accident (large pipe break) in which only the activity initially contained within the gap between the fuel pellet and cladding would be released into the containment. The core would not melt. It is assumed that the minimum required engineered safeguards would function satisfactorily to remove heat from the core and containment. The release would occur over the 0.5-hour period during which the containment pressure would be above ambient. Approximately 0.00001% of the iodines and 0.00006% of the alkali metals would be released. As in PWR category 8, the energy release would be very low.

BWR 1

This release category is representative of a core meltdown followed by a steam explosion in the reactor vessel. The latter would cause the release of a substantial quantity of radioactive material to the atmosphere. The total release would contain approximately 40% of the iodines and alkali metals present in the core at the time of containment failure. Most of the release would occur over a 1/2 hour period. Because of the energy generated in the steam explosion, this category would be characterized by a relatively high rate of energy release to the atmosphere. This category also includes certain sequences that involve overpressure failure of the containment prior to the occurrence of core melting and steam explosion. In these sequences, the rate of energy release would be somewhat smaller than for those discussed above, although it would still be relatively high.

BWR 2

This release category is representative of a core meltdown resulting from a transient event in which decay-heat-removal systems are assumed to fail. Containment overpressure failure would result, and core melting would follow. Most of the release would occur over a period of about three hours. The containment failure would be such that radioactivity would be released directly to the atmosphere without significant retention of fission products. This category involves a relatively high rate of energy release due to the sweeping action of the gases generated by the molten mass. Approximately 90% of the iodines and 50% of the alkali metals present in the core would be released to the atmosphere.

BWR 3
This release category represents a core meltdown caused by a transient event accompanied by a failure to scram or failure to remove decay heat. Containment failure would occur either before core melt or as a result of gases generated during the interaction of the molten fuel with concrete after reactor-vessel melt-through. Some fission-product retention would occur either in the suppression pool or the reactor building prior to release to the atmosphere. Most of the release would occur over a period of about 3 hours and would involve 10% of the iodines and 10% of the alkali metals. For those sequences in which the containment would fail due to overpressure after core melt, the rate of energy release to the atmosphere would be relatively high. For those sequences in which overpressure failure would occur before core melt, the energy release rate would be somewhat smaller although still moderately high.

BWR 4
This release category is representative of a core meltdown with enough containment leakage to the reactor building to prevent containment failure by overpressure. The quantity of radioactivity released to the atmosphere would be significantly reduced by normal ventilation paths in the reactor building and potential mitigation by the second containment filter systems. Condensation in the containment and the action of the standby gas treatment system on the releases would also lead to a low rate of energy release. The radioactive material would be released from the reactor building or the stack at an elevated level. Most of the release would occur over a 2-hour period and would involve approximately 0.08% of the iodines and 0.5% of the alkali metals.

BWR 5
This category approximates a BWR design basis accident (large pipe break) in which only the activity initially contained within the gap between the fuel pellet and cladding would not melt, and containment leakage would be small. It is assumed that the minimum required engineered safeguards would function satisfactorily. The release would be filtered and pass through the elevated stack. It would occur over a period of about 5 hours while the containment is pressurized above ambient and would involve approximately $6 \times 10^{-9}\%$ of the iodines and $4 \times 10^{-7}\%$ of the alkali metals. Since core melt would not occur and containment heat-removal systems would operate, the release to the atmosphere would involve a negligibly small amount of thermal energy.[15]

Table 2-4
Sabotage-Induced LOCA at PWR

Event Description	Containment Failure Mode[a]	Release Category
Large LOCA[b], ECCS[c]	ε	PWR 7
function failure	α	PWR 4
Large LOCA, ECCS function	β	PWR 5
failure, containment leakage	α	PWR 3
Large LOCA, ECCS	ε	PWR 7
injection failure	α	PWR 3
Large LOCA, ECCS injection	β	PWR 5
failure, containment leakage	α	PWR 3
Large LOCA, Electrical	ε	PWR 7
power disabled	γ	PWR 2
	δ	PWR 6
	α	PWR 1
Large LOCA, electrical power	β	PWR 2
disabled, containment leakage	α	PWR 1
Small LOCA, ECCS	β	PWR 7
injection failure	α	PWR 3
Interfacing systems LOCA	β	PWR 2
Large LOCA, containment	δ	PWR 3
heat removal failure	α	PWR 1

Source: Dean C. Kaul and Edward S. Sachs, *Adversary Actions in the Nuclear Fuel Cycle: I, Reference Events and Their Consequences*, SAI-121-612-7803 (Schaumburg, Ill.: Science Applications, 1977), table C-1. Reprinted with permission.

[a]Listed in order of decreasing likelihood: α, containment rupture due to a reactor vessel steam explosion; β, containment failure resulting from inadequate isolation of containment openings and penetrations; γ, containment failure due to hydrogen burning; δ, containment failure due to overpressure; ε, containment vessel melt-through.

[b]LOCA—loss of coolant accident.

[c]ECCS—emergency core cooling system.

Consequences of Damaged Reactors

Released aerosols are subject to complex processes in their distribution. Initially the cloud or plume is limited in size and heavily concentrated. According to one source, "The effect of the release at near downwind locations may be quite critically affected by minor topographical features (i.e., building, hills, and trees), by minor fluctuations in meteorological variables (i.e., wind direction and wind speed) and by release parameters (i.e., finite size of source and rate of release)."[16] Topographical features create a variety of local circulation patterns affecting both horizontal and vertical air

Table 2-5
Sabotage-Induced Transient at PWR

Event Description	· Containment Failure Mode[a]	Release Category
Transient and failure of power conversion and auxiliary feedwater systems	ϵ	PWR 7
	α	PWR 3
Transient, failure of power conversion and auxiliary feedwater systems, and containment leakage	β	PWR 5
	α	PWR 3
Transient, failure of power conversion and auxiliary feedwater systems, and electrical power failure	δ	PWR 2
	γ	PWR 2
	ϵ	PWR 6
	α	PWR 1
Transient, failure of power conversion and auxiliary feedwater systems, electrical power failure, and containment leakage	β	PWR 2
	α	PWR 1

Source: Kaul and Sachs, *Adversary Actions in the Nuclear Fuel Cycle*, table C-2. Reprinted with permission.
[a]Listed in order of decreasing likelihood: α, containment rupture due to a reactor vessel steam explosion; β, containment failure resulting from inadequate isolation of containment openings and penetrations; γ, containment failure due to hydrogen burning; δ, containment failure due to overpressure; and ϵ, containment vessel melt-through.

patterns which may vary during the day or evening. Weather is a major variable, with six categories of turbulence distinguishable: (A) very unstable, (B) moderately unstable, (C) slightly unstable, (D) neutral, (E) slightly stable, and (F) very stable.[17] The greater the stability, the more concentrated the released materials and the greater their lethal range for causing early fatalities. Greater turbulence disperses the plume but also broadens contamination. Regularities in weather patterns vary from site to site. For many locations, weather variability makes long-range prediction unreliable. Specific predictions for many locations cannot be made with great accuracy. However, as a general rule, neutral stability (category D) and inversions (categories E and F) characterize many sites 50 percent of the time, being particularly common in the evenings. Moisture further complicates the weather variables. Rain occurring at the time of a release concentrates fallout in the vicinity of the structure. Material carried by the rain-bearing clouds creates local hotspots. There may be early fatalities within an area of up to 150 mi. Rain occurring along the course of the cloud is not as effective in washing out material because the farther the cloud is from its source, the more diluted it becomes due to atmospheric mixing. In addition,

Table 2-6
Sabotage-Induced LOCA at BWR

Event Description	Containment Failure Mode[a]	Release Category
Large LOCA, ECCS	γ	BWR 3
function failure	β	BWR 2
	α	BWR 1
Large LOCA, ECCS	ζ	BWR 4
function failure,	α	BWR 1
containment leakage	β	BWR 2
Large LOCA, ECCS	γ	BWR 3
injection failure	γ'	BWR 2
	β	BWR 2
	α	BWR 1
Large LOCA, ECCS	$\delta\zeta$	BWR 3
injection failure,	β	BWR 2
containment leakage	α	BWR 1
Large LOCA, high	γ	BWR 3
pressure service water	γ'	BWR 2
system failure	α	BWR 1

Source: Kaul and Sachs, *Adversary Actions in the Nuclear Fuel Cycle*, table C-3. Reprinted with permission.

[a]Listed in order of decreasing likelihood: α, containment failure due to steam explosion in vessel; β, containment failure due to steam explosion in containment; γ, containment failure due to overpressure—release through reactor building; γ', containment failure due to over-pressure—release direct to atmosphere; δ, containment isolation failure in drywell; ζ, containment leakage greater than 2,400 volume percent per day.

ground runoff from rain dilutes material that is already deposited.[18] As the plume moves farther from the nuclear installation and diffuses horizontally and vertically, its trajectory increasingly will be subject to prevailing geostrophic winds (rather than local winds), dominating particulates reaching 600 to 900 m. However, such domination may not occur for many tens of miles.[19]

Release parameters include speed, the cloud's release height, composition, and deposition rate. Instantaneous releases result in narrower plumes than those over minutes or hours. Variable heights result from the plume's heat, weather conditions, and whether the release occurs at the top or bottom of a structure (the greater the release height, the more distant the extent of the plume). Plume composition varies depending upon the initial composition of the reactor core and the length of time it has been irradiated. Cores with fresh fuel contain fewer fission products than do cores that have been operating for some time. The deposition rate is important in

Table 2-7
Sabotage-Induced Transient at BWR

Event Description	Containment Failure Mode[a]	Release Category
Transient, failure	γ	BWR 3
to provide core	γ'	BWR 2
make-up water	α	BWR 1
Transient, failure		
to provide core	$\delta\zeta$	BWR 5
make-up water,	ζ	BWR 3
containment leakage	α	BWR 1
Transient, failure to	γ'	BWR 3
remove residual heat	γ	BWR 2
	α	BWR 1

Source: Kaul and Sachs, *Adversary Actions in the Nuclear Fuel Cycle*, table C-4. Reprinted with permission.

[a]Listed in order of decreasing likelihood: α, containment failure due to steam explosion in vessel; β, containment failure due to steam explosion in containment; γ, containment failure due to overpressure—release through reactor building; γ', containment failure due to overpressure—release direct to atmosphere; δ, containment isolation failure in drywell; ζ, containment leakage greater than 2,400 volume percent per day.

determining the concentration of material. This variable is assigned an uncertainty factor of 100 by the *Reactor Safety Study*.[20]

Each of these variables complicates consequence calculations, and methodological imperfections add to the problem. It is with these uncertainties in mind that the estimates should be reviewed. Calculations from several sources are drawn upon although the bulk of the data derives from little-publicized and in some cases previously unpublished work conducted at Princeton University, the most comprehensive analyses I could find. Because the calculations reflect a spectrum of assumptions, I have reproduced relevant charts and graphs with the text summarizing the most salient data. Although the estimates are based on accidents, the same results can occur through the destruction of a single nuclear reactor by conventional munitions or sabotage that takes advantage of facility vulnerabilities. The estimates do not reflect the destruction of several facilities commonly situated or the effects of nuclear weapons bombardment, which will be treated separately.

Figure 2-5 illustrates early fatality probabilities for a ground-level BWR 2 release for a 1,000 Mw(e) reactor assuming no protective action for twenty-four hours. The figure distinguishes the impact of two weather conditions. In turbulence category A (dashed line), the plume disperses quickly and has a lethal range of three miles, whereas in stable conditions (category

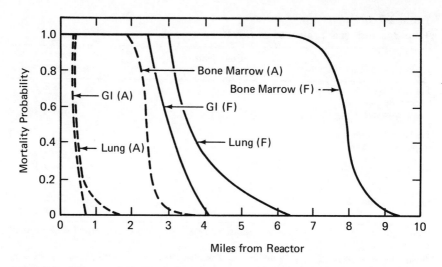

Sources: Nuclear Regulatory Commission, *Reactor Safety Study*, (Springfield, Va.: National Technical Information Service), Appendix VI, p. 13-9. (For other relevant curves see pp. 13-32, 13-33.) Information on the size of the reactor and release and length of exposure was obtained in a private communication from Rodger Blond of the Nuclear Regulatory Commission on May 1, 1980.

Note: Mortality probability for an affected population versus distance assuming no protective action for twenty-four hours from a 1,000 MW(e) reactor undergoing a BWR 2 in two hypothetical weathers. Stability category A, wind speed = 0.5 m/sec; stability category F, wind speed = 2.0 m/sec.

Figure 2-5. Early Mortality Curve

F), fatalities may occur up to 9 mi. where the plume may be 1 mi. wide. Both cases demonstrate the lethal prominence of bone marrow irradiation. Figure 2-6 from a Sandia Laboratories study indicates that late effects from major releases into the atmosphere can affect the most sensitive populations, fetuses and children, beyond 100 mi.

Jan Beyea, a physicist with Princeton University's Center for Environmental Studies, elaborated these findings in testimony to the New York State Board on Electric Generation Siting and the Environment concerning a 1,000 MW(e) reactor planned for Jamesport, Long Island, undergoing a PWR 2. Tables 2-8 to 2-14 present his calculations. They indicate the average rem dose for twenty-four hours at different distances and the probability of early death for one-day and one-week exposures with minimal and supportive medical treatment. The seven-day scenario is included to underscore the importance of evacuation and to take into account the possibility and Beyea's consideration that a stubborn minority may

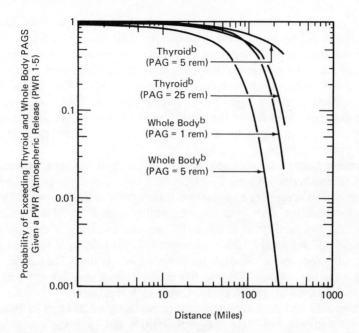

Source: D.C. Aldrich et al., *Examination of Offsite Radiological Emergency Measures for Nuclear Reactor Accidents Involving Core Melts*, SAND 78-0454 (Albuquerque, N.M.: Sandia Laboratories, 1978), p. 82. Reprinted with permission.

[a]Probabilities are conditional on a PWR atomspheric release (1-5). Shielding factor for airborne radionuclides = 1.0. Shielding factor for radionuclides deposited on ground = 0.7. One-day exposure to radionuclides on ground. Protective action guides are U.S. Environmental Protection Agency regulations establishing exposures warranting initiation of emergency protective actions.

[b]Whole body (thyroid) dose calculated includes external dose to the whole body (thyroid) due to the passing cloud and one-day exposure to radionuclides on ground, and the dose to the whole body (thyroid) from inhaled radionuclides within one year.

Figure 2-6. Conditional Probability of Exceeding Thyroid and Whole Body Protective Action Guides (PAGs) versus Distance for an Individual Located Outdoors[a]

refuse to evacuate or in a wartime scenario may be unable to do so due to military action. The focal point is F, E, and D weather stability, which characterizes Jamesport and assumes dry deposition (no rain).[21] Release heights are calculated for plume rises ranging from 0 to 1,000 m.

The calculations suggest several conclusions. The probability of death markedly increases with proximity to the reactor. Of greatest concern are twenty-four-hour exposures within 15 mi. although early fatalities can occur beyond 40 mi. in the most stable weather assuming relocation after seven days.[a] Thus countries (such as the Soviet Union) that build nuclear power plants in populated areas are exposing their citizens to a greater risk than those that do not. Weather instability diminishes the risk. For example, 10 mi. from a reactor, the probability of fatalities after a seven-day exposure is 45 percent, 68 percent, and 33 percent in stability classes F, E, and D respectively; it is 7 percent in stability class A. Unfortunately with the exception of a very few of the world's regions, such instability cannot be counted on to occur regularly. It would be imprudent for civil defense planning to hope otherwise.

Weather cannot be counted on to reduce fatalities, but evacuation, sheltering, and supportive medical treatment can. Calculations for one-day as compared to seven-day exposures call particular attention to the importance of evacuation. For example, in stability class E at 10 mi. from the reactor, the probability of death is 68 percent assuming relocation after seven days and minimal health treatment. This probability is reduced to 15 percent if exposure is limited to twenty-four hours. The tables also underline the importance of supportive medical treatment. For example, the 15 percent figure falls to 0.6 percent with such treatment.

Tables 2-13 and 2-14, distinguished respectively by 200-and 33-ft. tower wind distributions, translate the probabilities into average estimates of fatalities. These calculations suggest that with variable population densities projected for Jamesport in the year 2020 of roughly 120 to 1,000 persons per square mile, early deaths in stable (conditions D, E, and F) weather would range from 180 to 1,000 assuming evacuation and 390 to 1,700 with minimal treatment. If relocation of people took a week, fatalities could rise to 4,000 with supportive treatment and as high as 10,000 without such assistance. The greater release heights indicated in figure 2-13 are likely to increase the numbers of persons subject to lethal doses.

Figure 2-7 draws from a report produced jointly by the Nuclear Regulatory Commission and the Environmental Protection Agency, refines Beyea's findings through demonstration of the positive impact of rapidly implemented prophylactic measures in reducing fatalities. Presented are the mean number of projected early fatalities for PWR 1-5 assuming a uniform population density of 100 persons per square mile and irradiation from ground contamination. When exposures last twenty-four hours (bar 1) roughly seventy-five deaths are projected up to 25 mi. Sheltering in homes

---

[a]The area covered by the plume is distinct from the linear distance it reaches. Furthermore the area may vary considerably. Using sequences that maximized such problems as very severe contamination of the reactor site, lethal exposure at great distances, and maximum property damage, calculations for a 500 R inhaled whole-body dose using historical meteorological data from Oak Ridge, Tennessee, ranged from 9.8 km^2 to 149.2 km^2. C.V. Chester, "Estimates of Threat to the Public from Terrorist Acts Against Nuclear Facilities," mimeo. (Oak Ridge, Tenn.: Oak Ridge National Laboratory, n.d.), p. B-11.

Table 2-8
Average Dose Received after Spending Twenty-four Hours in Contaminated Region

Weather Stability Class	Wind Speed (mph)	Dose in Rem									
		5	10	15	20	25	30	35	40	45	50
F	4.5	547	214	119	77	54	41	31	25	20	17
	9										
	18										
	20										
E	4.5	613	191	89	51	32	22				
	9	416	150	76	46	31	22				
	18	250	100	56	35	24	18				
	20										
D	4.5	217	95	57	39	28	21	16	12	10	8
	9	150	67	43	30	22	17	13	11		
	18	95	45	29	21	16	12	10	8		
	20										
C[a+]	4.5	110	48	28	19	13					
	9										
	18										
	20										
B[a+]	4.5	84	38	23	15	11					
	9	50	24	15	11	8					
	18	27	14	9	7	5					
	20										
A[a+]	4.5	70	31	18	12	8					
	9	40	19	12	9	6					
	18	21	11	7	5	4					
	20										

Source: Jan Beyea, "In the Matter of Long Island Lighting Company (Jamesport Nuclear Power Station, Units 1 and 2)," *Direct Testimony* of Dr. Jan Beyea (New York: New York State Board on Electric Generation Siting and the Environment, Case No. 80003, May 1977, mimeo.), table 6. For additional data, see Jan Beyea, "Program BADAC, Short-term Doses Following a Hypothetical Core Melt-down" (available from the author).

Note: Gaussian plume model, top-hat approximation, deposition included as in *Reactor Safety Study*. Plume rise and deposition velocity parameterized and resultant mortality probabilities averaged as follows: deposition velocity: uniform on log scale between 0.001 and 0.1 m/sec.; release height: 0-250 m for classes E and F, 0-600 m for class D, and 0-1,000 m for classes A, B, and C; mixing height: 1,000 m; shielding factors = 1/3 for ground dose, 3/4 for cloud dose; dose conversion factors and response factors from *Reactor Safety Study*, appendix VI.

[a+] Averages calculated for three plume rises: 0, 500, and 1,000 m.

Table 2-9
Average Probability of Death as a Function of Distance along Plume Path: Evacuation after Twenty-four Hours and Minimal Treatment

Weather Stability Class	Wind Speed (mph)	Miles from Reactor									
		5	10	15	20	25	30	35	40	45	50
F	4.5	38%	31%	11%	1.4%	0.03%					
	9										
	18										
	20										
E	4.5	69	15	.05							
	9	48	10	.045							
	18	30	3	.009							
	20										
D	4.5	26	6	.0006							
	9	18	3	.0002							
	18	11	.6	.0002							
	20										
C	4.5										
	9										
	18										
	20										
B	4.5										
	9										
	18										
	20										
A	4.5										
	9										
	18										
	20										

Source: Beyea, "In the Matter of Long Island Lighting Company," table 3c.
Note: Gaussian plume model, top-hat approximation, deposition included as in *Reactor Safety Study*. Plume rise and deposition velocity parameterized and resultant mortality probabilities averaged as follows: deposition velocity: uniform on log scale between 0.001 and 0.1 m/sec.; release height: 0–250 m for classes E and F, 0–600 m for class D, and 0–1,000 m for classes A, B, and C; mixing height: 1,000 m; shielding factors = 1/3 for ground dose, 3/4 for cloud dose; dose conversion factors and response factors from *Reactor Safety Study*, appendix VI.

Table 2-10
Average Probability of Death as a Function of Distance along Plume Path: Evacuation after Twenty-four Hours and Supportive Treatment

Weather Stability Class	Wind Speed (mph)	Miles from Reactor									
		5	10	15	20	25	30	35	40	45	50
F	4.5	34%	17%	1%	0.0003%						
	9										
	18										
	20										
E	4.5	52	.6								
	9	34	.2								
	18	17	.0006								
	20										
D	4.5	16	.6								
	9	10	.02								
	18	5									
	20										
C	4.5										
	9										
	18										
	20										
B	4.5										
	9										
	18										
	20										
A	4.5										
	9										
	18										
	20										

Source: Beyea, "In the Matter of Long Island Lighting Company," table 3d.

Note: Gaussian plume model, top-hat approximation, deposition included as in *Reactor Safety Study*. Plume rise and deposition velocity parameterized and resultant mortality probabilities averaged as follows: deposition velocity: uniform on log scale between 0.001 and 0.1 m/sec.; release height: 0-250 m for classes E and F, 0-600 m for class D, and 0-1.000 m for classes A. B, and C; mixing height; 1,000 m; shielding factors = 1/3 ground dose, 3/4 for cloud dose; dose conversion factors and response factors from *Reactor Safety Study*, appendix VI.

Table 2-11
Average Probability of Death as a Function of Distance along Plume Path: Relocation after Seven Days and Supportive Treatment

Weather Stability Class	Wind Speed (mph)	Miles from Reactor									
		5	10	15	20	25	30	35	40	45	50
F	4.5	43%	41%	32%	16%	6%	1%	0.02%			
	9										
	18										
	20										
E	4.5	81	51	9	.002						
	9	63	40	8	.003						
	18	46	24	2.5	.0006						
	20										
D	4.5	34	21	8	.8	.0007					
	9	26	14	4	.03						
	18	18	8	.5							
	20										
C	4.5										
	9										
	18										
	20										
B	4.5										
	9										
	18										
	20										
A	4.5	20	.007								
	9										
	18										
	20										

Source: Beyea, "In the Matter of Long Island Lighting Company," table 3b.

Note: Gaussian plume model, top-hat approximation, deposition included as in *Reactor Safety Study*. Plume rise and deposition velocity parameterized and resultant mortality probabilities averaged as follows: deposition velocity: uniform on log scale between 0.001 and 0.1 m/sec.; release height: 0-250 m for classes E and F, 0-600 m for class D, and 0-1,000 m for classes A, B, and C; mixing height; 1,000 m; shielding factors = 1/3 ground dose, 3/4 for cloud dose; dose conversion factors and response factors from *Reactor Safety Study*, appendix VI.

Table 2-12
Average Probability of Death as a Function of Distance along Plume Path: Relocation after Seven Days and Minimal Health Treatment

Weather Stability Class	Wind Speed (mph)	Miles from Reactor									
		5	10	15	20	25	30	35	40	45	50
F	4.5	46%	46%	42%	32%	20%	10%	4%	1.2%	0.3%	0.03%
	9										
	18										
	20										
E	4.5	93	68	36	6	.17	.0016				
	9	76	53	30	6	.2	.002				
	18	58	37	19	3	.002					
	20										
D	4.5	40	33	20	9	2	.3	.02	.0001		
	9	33	25	13	4	.07	.03	.0008	.0003		
	18	23	16	6	1	.07	.0007	.0002			
	20										
C	4.5										
	9										
	18										
	20										
B	4.5										
	9										
	18										
	20										
A	4.5	32	7	.07							
	9										
	18										
	20										

Source: Beyea, "In the Matter of Long Island Lighting Company," table 3a.

Note: Gaussian plume model, top-hat approximation, deposition included as in *Reactor Safety Study*. Plume rise and deposition velocity parameterized and resultant mortality probabilities averaged as follows: deposition velocity: uniform on log scale between 0.001 and 0.1 m/sec.; release height: 0-250 m for classes E and F, 0-600 m for class D, and 0-1,000 m for classes A, B, and C; mixing height; 1,000 m; shielding factors = 1/3 ground dose, 3/4 for cloud dose; dose conversion factors and response factors from *Reactor Safety Study*, appendix VI.

The Vulnerability of Nuclear Facilities   41

Table 2-13
Average Number of Early Deaths to be Expected from a PWR 2 Accident at Jamesport in the Year 2020: Two-Hundred-Foot Tower Wind Distributions

Weather Stability Class	Wind Speed (mph)	Relocation after 7 Days		Evacuation in 24 Hours	
		Minimal Treatment	Supportive Treatment[a]	Minimal Treatment	Supportive Treatment[a]
F[c]	4.5	10,000[b]	4,000	1,300	680
	9	2,300	1,300	630[g]	350[g]
	18	1,700	950[g]	400[g]	200[g]
	20				
E[c]	4.5	5,000	3,100	1,700	1,000
	9	4,000	2,300	1,100	640
	18	2,700	1,600	670	340
	20	2,300[h]			
D[d]	4.5	4,200	2,000	890	440
	9	2,900	1,400	600	280
	18	1,900	950	390	180
	20				
C	4.5				
	9				
	18				
	20				
B	4.5				
	9				
	18				
	20				
A[e]	4.5				
	9				
	18				
	20				
Averaged over all classes and speeds[f]		1,700	980	430	220

Source: Beyea, "In the Matter of Long Island Lighting Company."

Note: Average number of deaths means an average over many hypothetical accidents at the same site. For population data, see table I of Beyea's testimony. Early deaths assumes within sixty days. Dose reponse curves from *Reactor Safety Study*, appendix VI, fig. 9-1.

[a]Definition of supportive treatment "indicates such procedures as reverse isolation . . . , sterilization of all objects in patient's room, use of . . . laminar-air-flow systems, large doses of antibiotics, and transfusions of whole blood packed cells or platelets." Ibid., p. F-1

[b]Reduced about 25 percent if Connecticut excluded.

[c]Mortality probabilities used to generate these numbers were calculated by assuming random effective release height (between ground and 250 m), random deposition velocity (on log scale) between 0.001 and 0.1 m/sec., and time invariant weather.

[d]Same as note c except that release height range taken between ground and 600 m.

[e]Same as note c except that release height range taken between ground and 1,000 m.

[f]Numbers will go up slightly when A, B, and C classes are included.

[g]Determined using E mortality probabilities.

[h]Determined using 18 mph mortality probabilities.

Table 2-14
Average Number of Early Deaths to be Expected from a PWR 2 Accident at Jamesport in the Year 2020: Thirty-Three-Foot Tower Wind Distributions

Weather Stability Class	Wind Speed (mph)	Relocation after 7 Days		Evacuation in 24 Hours	
		Minimal Treatment	Supportive Treatment[a]	Minimal Treatment	Supportive Treatment[a]
F[d]	4.5	6,200[b]	2,700[c]	1,000	590
	9				
	18				
	20				
E[d]	4.5	4,800	3,000	1,600	1,000
	9	3,700	2,300	1,100	650
	18	2,200	1,300	550	280
	20				
D[e]	4.5	3,800[c]	1,700	770	380
	9	3,100	1,500	670	310
	18	2,000	1,000	430	190
C	4.5				
	9				
	18				
	20				
B	4.5				
	9				
	18				
	20				
A[f]	4.5				
	9				
	18				
	20				
Averaged over all classes and speeds[g]		2,200	1,300	590	320

Source: Beyea, "In the Matter of Long Island Lighting Company."

Note: Average number of deaths means an average over many hypothetical accidents at the same site. For population data, see table I of Beyea's testimony. Early deaths assumes within sixty days. Dose response curves from *Reactor Safety Study*, appendix VI, fig. 9-1.

[a]Definition of supportive treatment "indicates such procedures as reverse isolation . . . , sterilization of all objects in patient's room, use of . . . laminar-air-flow systems, large doses of antibiotics, and transfusions of whole blood packed cells or platelets." Ibid., p. F-1

[b]Reduced about 40 percent if Connecticut excluded.

[c]Reduced about 15 percent if Connecticut excluded.

[d]Mortality probabilities used to generate these numbers were calculated by assuming random effective release height (between ground and 250 m), random deposition velocity (on log scale) between 0.001 and 0.1 m/sec., and time invariant weather.

[e]Same as note d except that release height range taken between ground and 600 m.

[f]Same as note d except that release height range taken between ground and 1,000 m.

[g]Numbers will go up slightly when A, B, and C classes are included.

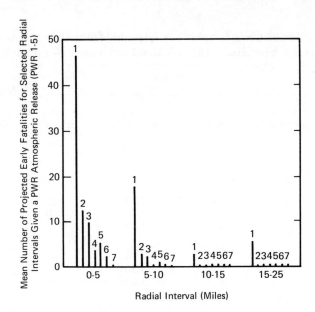

Source: U.S. Nuclear Regulatory Commission and U.S. Environmental Protection Agency Task Force on Emergency Planning, *Planning Basis for the Development of State and Local Government Radiological Emergency Response Plans in Support of Light Water Nuclear Power Plants*, NUREG-0396, EPA 520/1-78-016 (Springfield, Va.: National Technical Information Service, December 1978).

Notes: (1) No immediate protective action, shielding factors (SFs)[a] (0.75, 0.33),[b] one-day exposure to radionuclides on ground. (2) Sheltering, SFs (0.75, 0.33), six-hour exposure to radionuclides on ground. (3) Sheltering, SFs (0.5, 0.08), one-day exposure to radionuclides on ground. (4) Sheltering, SFs (0.5, 0.08), six-hour exposure to radionuclides on ground. (5) Evacuation, five-hour delay time, 10 miles per hour (mph). (6) Evacuation, three-hour delay time, 10 mph. (7) Evacuation, one-hour delay time, 10 mph. Assumes a uniform population density of 100 persons per mi.[2]

[a]Shielding factors (airborne radionuclides, ground contamination).

[b]Shielding factors for no protective action were chosen to be the same for sheltering in areas where most homes do not have basements.

Figure 2-7. Mean Number of Projected Early Fatalities within Selected Radial Intervals for Evacuation and Sheltering Strategies Given PWR Atmospheric Release (PWR 1-5)

without basements "with 6 hours of effective exposure to ground contamination" (bar 2) reduces this number to about sixteen. Four or five people (bar 4) might die in homes with basements. Rapid evacuation (bars 6 and 7) reduces this figure still further. In sum, these and the previous calculations suggest that the early consequences of releases can vary enormously depending upon the mix of variables.

Late effects such as cancers, thyroid damage, and genetic defects can vary as well; however the numbers of casualties may be significantly higher than early fatalities. As the radioactive plume disperses, it will spread in a wedge shape with its horizontal and vertical extensiveness determined by turbulence. Figure 2-8 distinguishes the greater concentration but lesser horizontal breadth of a clear evening release compared to a more turbulent clear day release. The cloud, ever widening, can extend great distances, taking from one-half hour to three hours to pass a point. Along the way, it will deposit products and, assuming no evauation, irradiate an increasingly larger number of people. The degree of deposition varies. For example, a major release in evening overcast forms a wedge 12.5 mi. in breadth that exposes populations from 0 to 400 rem; at 150 mi. this wedge is 18.6 mi., exposing populations to 0 to 200 rem.[22]

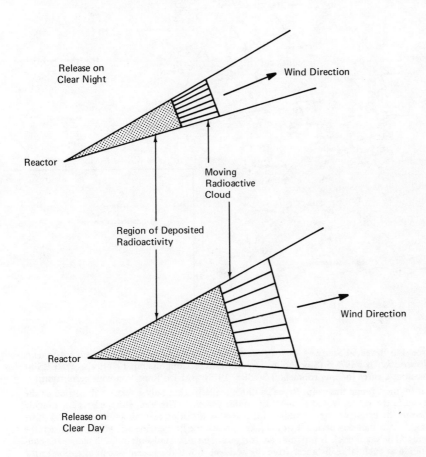

Source: Jan Beyea, "The Impact in New York City of Reactor Accidents at Indian Point," statement to the New York City Council, June 11, 1979, corrected June 20, 1979 (Princeton, N.J.: Center for Energy and Environmental Studies, 1979), p. 30.

Figure 2-8. Top View of Plume

Beyea elaborates upon these consequences in the study of a 580 MW(e) Swedish reactor that, at this output, generates less than 60 percent of Jamesport's production. The plant (actually there are two reactors, each the same size) is located in the vicinity of Barsebäck, a coastal village fifteen miles east of Copenhagen and twelve miles north of Malmö (figure 2-9). Beyea ran one thousand computer accident simulations varying weather,

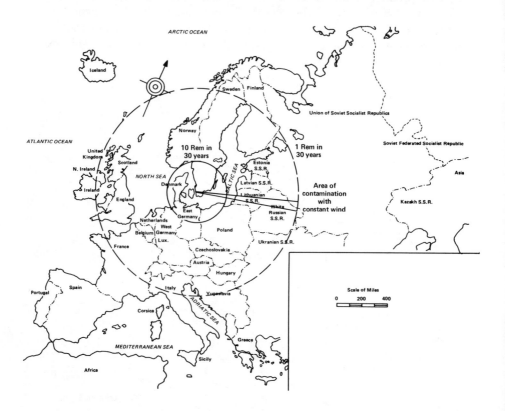

Source: Jan Beyea, *A Study of Some of the Consequences of Hypothetical Reactor Accidents at Barsebäck*, Ds I 1978:5 (Stockholm: Swedish Energy Commission, 1978); also Pu/CES 61 (Princeton: Center for Environmental Studies, 1978), p. I-15. Reprinted with permission.

Note: In interpreting this map, Beyea cautions readers "that only a very small faction of the land inside the circles would actually be contaminated." The wedge-shaped region roughly represents the extent of contamination for one wind direction assuming no major wind shifts during the one half-day period it would take to reach the first circle and two days to reach the second. "The outer circle is probably an exaggeration since wind shifts should reduce the concentration at such large distances from the accident, but it does serve to call attention to the potential international impact of a catastrophic accident or a major release resulting from conventional weapons destruction occurring in Europe." Ibid., p. I-13.

Figure 2-9. Barsebäck Distances of Concern for Land Contamination

wind direction, and speed for BWR 1, 2, and 3 releases. Concluding that the effect of weather is less significant for long-term consequences than for short-term ones, he tracked the approximate breadth of radioactivity 1,000 km from Barsebäck over a region with an average population density of 50/km^2. The average exposed population was one million.

Table 2-15 tabulates the breadth of contamination in stability class D with a 7.5 m/sec. wind for a BWR 2. The calculations illustrate the distances and area of land contamination of 10 rem in thirty years for rural land and 25 rem in thirty years for urban land that the *Reactor Safety Study* considers unsafe for occupancy and a more conservative 1 rem criterion in

Table 2-15
Land Contamination for a BWR 2 Accident during Typical Barsebäck Weather
(in km from reactor)

	Distance to Which Land Is Contaminated above Interdiction Threshold[c]		Contaminated Area (including Water)	
	Average	Maximum	Average	Maximum
Reactor Safety Study criteria				
Rural land criterion[a]	380 km	600 km	10,000 km^2	24,000 km^2
If population kept out for 6 months	300	480	6,200	15,000
If population kept out for 10 years	220	300	3,700	6,500
Urban land criterion[b]	230	380		
1 Rem in 30 years criterion	1,300	2,000	100,000	190,000
If population kept out for 6 months	1,000	1,600	64,000	145,000
If population kept out for 10 years	750	1,150	35,000	80,000

Source: Jan Beyea, *A Study of Consequences of Hypothetical Reactor Accidents at Barsebäck*, Ds I 1978:5 (Stockholm: Swedish Energy Commission, 1978), and PU/CES 61 (Princeton: Center for Environmental Studies, 1978), p. I-14. Reprinted with permission.

Note: Wind 7.5 m/sec., stability class D. "Average" means over and plum rise. "Maximum" means maximum contamination as deposition and plume height are varied.

[a]10 rem in thirty years.

[b]25 rem in thirty years. The reduction in dose due to cesium migration into soil is assumed equal to the effect of reduced shielding in urban areas and cesium wash-off.

[c]Distance from the reactor.

thirty years for both. The urban-rural dichotomy presumes that the expense and risk in urban relocation outweigh the reduction in exposure below 25 rem.[23] Both maximum and average findings are included. They suggest that at the very worst 24,000 km² would be contaminated on the basis of the *Reactor Safety Study* criteria and 190,000 km² using the more conservative measure. The results also indicate how radiation lingers for many years, although it decreases most rapidly at its outermost reaches. For example, the average distance in rural areas land is contaminated above the interdiction threshold, using the *Reactor Safety Study*'s criteria, declines from 380 km to 300 km in six months and 220 km in ten years.

Table 2-16 presents Beyea's casualty findings averaged using the linear

Table 2-16
Thyroid Nodule Cases and Cancer Deaths at All Distances from Barsebäck
(averaged over 1,000 accident simulations)

	Total Person-Rem[a]		
Organ	*BWR 1*	*BWR 2*	*BWR 3*
Thyroid[b]	2.3×10^8	3.3×10^8	$.5 \times 10^8$
Whole body dose[c]	2×10^7	1.1×10^7	$.5 \times 10^7$
Long-term dose from ground contamination[e]	$0\text{-}7.3 \times 10^7$	$0\text{-}8.4 \times 10^7$	$0\text{-}1.8 \times 10^7$
Cancers			
Thyroid nodules[d]	62,000-300,000	90,000-425,000	14,000-65,000
Thyroid cancer	3,000-18,000	4,000-25,000	600-3,700
(over 20-year period)	(450-2,500 fatal)	(600-3,700 fatal)	(100-500 fatal)
Other cancer deaths from whole body dose within 1 week[c]	2,600	1,400	600
Cancer deaths from long-term ground contamination[e]	0-9,500	0-11,000	0-2,400
Total fatal cancers	3,000-15,000	2,000-16,000	700-3,500

Source: Beyea, *A Study of the Consequences.* p. I-8.

Note: The average exposed population is about 1 million, average population density about 50/km². Cancer calculations use linear hypothesis and coefficients from Appendix.

[a]Note that, unlike the *Reactor Safety Study*, no dose reduction factors were used for low individual doses. If they had been used, the whole body dose would be two times lower.

[b]No dose reduction factors were used for I^{131}. A range of dose reduction factors of 0.1-1 is reflected in the range of exposure-dose coefficients given in the Appendix.

[c]Following the *Reactor Safety Study*, the table assumes twenty-four hours spent in contaminated ground close to the reactor (less than 50 km) and one week far from the reactor (greater than 50 km). See figure A-3 of Beyea's report for dose frequency distribution. Whole body dose set equal to marrow dose.

[d]The higher numbers given are so large compared to the total exposed population that saturation effects would reduce them.

[e]The larger number assumes no interdiction and no ground decontamination. Doses were calculated starting one week after the accident.

consequence hypothesis, which assumes "that a number of small doses given to many individuals or spread out in time have the same effect as a smaller number of larger doses, equal in the aggregate to the sum of the smaller ones, given to fewer individuals at one time."[24] This provides a rough approximation of the actual dose-effect relation, which will vary depending on the cancer. The computations illustrate that the dominant effect will be tens to hundreds of thousands of nonfatal thyroid nodules, although Beyea qualifies the upper figure in a footnote to his table: "The high numbers given are so large, compared to the total exposed population that saturation effects would reduce them." Fatalities range from 700 to 16,000 depending upon the magnitude of the release. The calculations averaged varying weather. Not shown in these figures are the upper ranges of consequences, which are found in 10 percent of the simulations that long-term health effects were approximately a factor of two or more higher than the average values, reflecting mainly winds blowing toward more populated regions in the Netherlands, Belgium, and Germany. In 1 percent of the simulations, the effects were three times as great as the averages.[25]

The computations suggest that the largest potential contributor to fatalities is long-term ground contamination. The average upper limit assumed no long-term land-use restrictions and no decontamination. This picture is not likely to be realistic, but it does point out a policy dilemma. Beyea notes, "It would be possible in principle to relocate people permanently. However, in practice, if the contaminated area should be large, policy makers would face difficult decisions in setting interdiction and decontamination thresholds. Large areas would be involved. Lowering the level of acceptable individual risk would raise the cost of interdiction and decontamination."[26] There are many uncertainties concerning costs, including those for the evacuation of populations to reduce exposure; temporary or permanent relocation of people who live in regions contaminated beyond an acceptable threshold; decontamination; and temporary or permanent denial of land, agriculture, dwellings, and factories. They will vary from country to country and, of course, depend on the extensiveness and intensity of contamination. In the United States, the authors of the *Reactor Safety Study* calculated that the costs could range from less than $1 million for minor releases (not including damage to the power plants or costs replacing power generation) to $14 billion for a major release.[27] Furthermore there are uncertainties about the effectiveness of decontamination, which includes removing the top layer of soil or overturning it and scraping and washing man-made structures. The *Reactor Safety Study*'s authors suggest that such methods can reduce contaminants by a factor of twenty, reflected in the calculations in table 2-17. However, other assessments are more cautious. According to the American Physical Society's study group on light-water reactors, "Thresholds and agricultural plans for eventual reoccupation of contaminated areas all suffer from a lack of definitive knowledge concerning the fixing of radionuclides in the soil, buildings, etc.,

Table 2-17
Effect of a Decontamination Factor of Twenty on the Land Contamination of Table 2-15
(in km from reactor)

	Distance to Which Land Is Contaminated above Interdiction Threshold[a]		Contaminated Area (including Water)	
	Average	*Maximum*	*Average*	*Maximum*
Reactor Safety Study criterion	80 km	140 km	325 km^2	1,100 km^2
1 rem in 30-year criterion	260	430	4,500	13,000

Source: Beyea, *Study of Some of the Consequences*, p. I-14.

and their subsequent history."[28] Medical examinations of Marshall Islanders who returned to Bikini Island in 1969, the site of a 1954 hydrogen weapons test, underscored this point when they revealed that the inhabitants were ingesting large quantities of cesium and strontium-90 by eating fruits grown on the island. Consequently the United States government evauated the island permanently.[29]

In addition to early and late somatic effects, Beyea calculated average genetic consequences over a five- to ten-generation period following inhalation, twenty-four-hour, and seven-day ground exposures at 50 km and over 50 km, respectively, and long-term ground contamination beginning seven days after the accident to infinity. Table 2-18 presents these estimates which range to ten generations following exposure. The consequences vary with the magnitude of the release from a few hundred genetic defects and spontaneous abortions to possibly several tens of thousands of cases of genetically related diseases. As in the case of late somatic defects, Beyea calculated that the figures doubled in 10 percent of the simulations and tripled in 1 percent.

There is reason to believe that both Beyea's Barsebäck and Jamesport calculations are not conservative from the perspective of wartime effects. This is true for the Swedish study because the reactor addressed is roughly 60 percent the size of many other plants. For example, compare this case to what would have happened at the Three Mile Island 880 MW(e) PWR had a major accident occurred. Table 2-19 presents the estimates. In a PWR 2 (TMI-5a,b) radiation that would linger for many years could cover from 1,400 mi.^{2} to 5,300 mi.^{2}, the larger figure reflecting a reactor core in operation for longer than three months (a mature core). (The table also indicates late fatalities for this release, ranging from several hundred up to 60,000, reflecting a larger exposed population than Barsebäck. Footnote c suggests that the estimates are mid-range; using the *Reactor Safety Study* conse-

Table 2-18
Barsebäck Consequences Resulting in Genetic Damage

Cases	BWR 1	BWR 2	BWR 3
From one-week exposure[a]			
Genetic defects[b]	500-5,000	200-2,500	100-1,000
Spontaneous abortions	850	450	200
Genetically related diseases[c]	0-10,000	0-5,000	0-2,500
From long-term ground contamination[d]			
Genetic defects[b]	0-18,000	0-21,000	0-4,500
Spontaneous abortions	0-3,000	0-3,500	0-750
Genetically related diseases[c]	0-36,000	0-42,000	0-9,000

Source: Beyea, *A Study of the Consequences*, p. I-9.

Note: Average exposed population is approximately 1 million, average population density is 50/km². Effects of doses are calculated using the linear hypothesis and coefficients from Appendix. Data are averaged over 1,000 accident simulations.

[a]From inhalation and cloud dose; twenty-four hour ground dose within 50 km, seven-day ground dose beyond 50 km.

[b]Persons with identifiable dominant genetic defects over an average of five generations following exposure.

[c]Total extra constitutionally or degeneratively diseased persons over an average of ten generations following exposure.

[d]From ground dose beginning seven days after accident to infinity.

quence model, it is high by a factor of four but low by an equal amount using more conservative models.) Furthermore neither the Jamesport nor the Swedish case addresses consequences of product release from several commonly situated reactors and spent fuel depositories. Many countries cluster their reactors. Nor do the calculations suggest the consequences from reactors, notably the LMFBR, which is fueled with 10 to 20 percent plutonium, which contain greater concentrations of long-lived toxic products. Finally the estimates are not necessarily applicable to destruction by nuclear weapons.

The consequences from atomic weapons are distinguished assuming that an accurately delivered explosive is burst near or on the ground (ground burst) fragmenting the pressure vessel and entraining the reactor's products in the stem of its cloud. A 100 kiloton weapon detonated approximately 200 ft. away or a 10 Mton weapon 2,000 ft. away could achieve a breach of the pressure vessel.[30] This explosion significantly extends the lethal range of the nuclear fallout estimated in the case of an LMFBR to be approximately 17 percent for a week's exposure and 33 percent for a month's exposure (figure 2-10). Such extension may likewise apply to LWRs although the increase is not provided in the available literature.

Table 2-19
Some Long-term Consequences of Hypothetical Accidents at Three Mile Island

Accident Designation	Releases to Atmosphere	Delayed Cancer Deaths[a,b] (low/high)[c]	Thyroid Nodule Cases[b,d] (low/high)	Temporary Agricultural Restrictions	Areas Requiring Decontamination or Long-Term Restrictions on Occupation[e]
TMI-0	10% of noble gases (similar to actual accident)	0/4	0	0	0
Releases greater than actually occurred					
TMI-1	60% of noble gases	1/25		0	0
TMI-2	5% iodines plus 60% noble gases	3/350	200/27,000	25,000 mi.²[f]	0
TMI-3a	TMI-2 plus 10% of cesiums	15/2,000	200/27,000	25,000 mi.²[f]	75 mi.²
TMI-4a	50% of cesiums	100/12,000		3,700 mi.²[f]	650 mi.²
TMI-5a	PWR 2 release with complete core melt[i]	200/23,000	3,500/450,000	175,000 mi.²[f]	1,400 mi.²
Consequences assuming reactor core had been in operation for much longer than 3 months (mature core)					
TMI-3b	TMI-2 plus 10% of cesiums	65/8,500	200/27,000	25,000 mi.²[f]	550 mi.²
TMI-4b	50% of cesiums	440/48,000[j]		18,000 mi.²[g]	4,300 mi.²
TMI-5b	PWR 2 release[h]	550/60,000[j]	3,500/450,000	175,000 mi.²[f]	5,300 mi.²

Source: Jan Beyea, "Some Long-Term Consequences of Hypothetical Major Releases of Radioactivity to the Atmosphere from Three Mile Island," draft (Princeton, N.J.: Center for Energy and Environmental Studies, September 7, 1979), p. 5. Reprinted with permission.

Note: All accidents are assumed to take place under typical meteorological conditions. Wind shifts and changes in weather neglected. Details can be found in supporting tables in appendixes B and E in ibid. Health effects are totaled for people living beyond 50 mi. Does not include any early illness or deaths that might be associated with high dose to unevacuated populations a few tens of miles from the reactor.

[a]Cumulative total over a seventy-five-year period after the accident. The range of genetic defects would be equal, very roughly, to the range of delayed cancer deaths.

[b]The low number is for the most favorable wind direction (eastern Maryland), assuming the most optimistic coefficient relating dose to health effects, and evacuation out to 50 mi. (Without evacuation, the low number would be a factor of two to five higher depending on the accident.) The high number is for the least favorable wind direction (New York City/Boston) and assuming the most pessimistic coefficient relating dose to health effects. (Evacuation also is assumed out to 50 mi. but has a small impact on the high results.)

[c]Reduce high value by a factor of about four to obtain the prediction that would result using the *Reactor Safety Study* model. Multiply by four to obtain the prediction that would result using health effects coefficients based on data of Mancuso, Stewart, and Kneale. See appendix E in Beyea's Three Mile Island report for elaboration.

[d]Cumulative total over a twenty-five-year period after the accident. A zero implies a small number.

[e]See table B-V in appendix B in Beyea's report for details.

[f]Milk restrictions. See table B-IV in Beyea's report for elaboration.

[g]First-year crop restrictions (harvested food not suitable for children).

[h]A PWR 2 accident as defined in the *Reactor Safety Study*.

[i]This number possibly could be reduced in half if massive decontamination or relocation efforts were undertaken in urban areas to avoid low-level radiation doses.

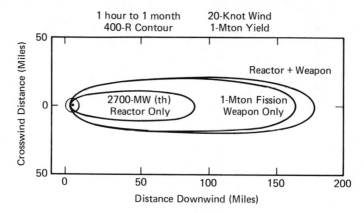

The 400-R isodose contours for one hour to one
month for fallout from a 1000-MW(e) reactor,
1-Mton fission weapon, and combination.

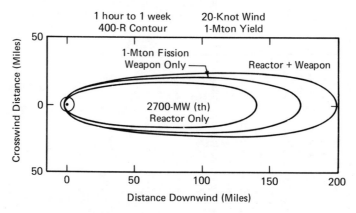

The 400-R isodose contours for one hour to one
week for fallout from a 1000-MW(e) reactor,
1-Mton fission weapon, and combination.

Source: Conrad V. Chester and Rowena O. Chester, "Civil Defense Implications of a LMFBR
in a Thermonuclear Target Area," *Nuclear Technology* 21 (March 1974):191. Reprinted with
permission.
Figure 2-10. 1,000 MW(e) LMFBR and 1-Mton Fission Weapon Contours

Nuclear explosives that are detonated at such a height that the fireball
does not touch the earth's surface (air burst) or ground burst up to 8 km for
a 1 Mton device will create overpressure sufficient to disrupt vital systems.
The result would be a meltdown, but the releases would not add significant
quantities outside the region depopulated by the weapon. However, the
reactors would add significant long-lived radioactivity to the environment

well beyond that produced by the weapon. Figure 2-11 demonstrates that the residual activity from a 100 kt weapon declines below that of a 1,000 MW(e) reactor in less than a day and continues to fall rapidly, while that of the power plant remains relatively high. The initial consequences of a 1 Mton nuclear weapon detonation with lethal effects extending beyond 150 mi. are many magnitudes greater than those from the reactor containing the most toxic concentrations of radioactive products.

Nuclear Reactor Support Facilities

Incentives to Destroy Atomic Power Plant Support Facilities

Nuclear support facilities may be less attractive military targets than are reactors for several reasons. First, they do not share the same strategic importance. Since the reactors can operate for months without refueling, their destruction cannot immediately cripple energy production. Second, with the exception of spent fuel pools located at every reactor site, they are less vulnerable by virtue of their limited numbers. Only ten countries—France, West Germany, India, Japan, England, the United States, the Soviet Union, Belgium, Argentina, and Brazil—operate or plan to fabricate mixed oxide fuel using plutonium. Commercial reprocessing plants that include spent fuel, high-level liquid waste, and plutonium storage facilities are currently limited to France and England. A third, in Belgium, is closed temporarily. Other such plants are planned for Germany, Italy, Japan, Spain, Pakistan, the Soviet Union, and India although there is considerable uncertainty in each case whether these projects will be built.[31] Finally, with the exception of spent fuel at reprocessing plants and high-level liquid waste storage, the amount of fission products and actinides that can be released, particularly by conventional explosives, will be less significant than the quantities from reactors.

Still, although less attractive than reactors, some of these facilities will be alluring. Each makes its own contribution to the fuel cycle. Reprocessing and fabrication plants have intrinsic economic value, and each could contribute to a weapons program. All such plants can release some nuclear products if subjected to explosives, and some enormous quantities. An acquaintance with the operation of these installations will help explain their vulnerability in war.

The Vulnerability of Support Facilities

Each year about one-third of a PWR core (29 MT) and one-fourth of a BWR core (37 MT) of spent fuel is discharged, to be replaced by fresh

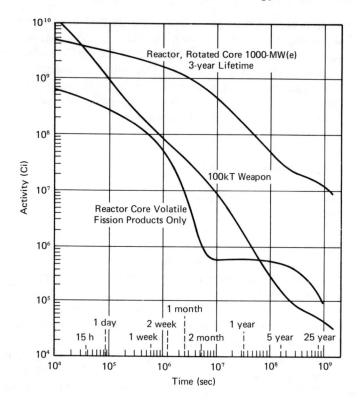

Source: Chester and Chester, "Civil Defense Implications," p. 786. Reprinted with permission.

Figure 2-11. Residual Activity From a 1,000 MW(e) Reactor and a 100-kT Weapon versus Time after Weapon Detonation or Reactor Shutdown

fuel.[32] The material is highly radioactive. In addition to unburned uranium, each ton contains about 30 kg of fission products, and slightly less than 10 kg of actinides emitting about 300 million ci of activity at the time of reactor shutdown.[33] The composition will vary depending upon the fuel in use and its burn-up, with longer burn-ups containing higher concentrations.

Because this material is so radioactive, it is very hot. One day after shutdown, 30 tons of spent LWR fuel has a thermal output of 10,000 kW. The output is even higher for some other reactor types, such as the breeder. To prevent the spent fuel from melting once it is removed from the pressure vessel, it is placed on storage racks in rectangular pools, typically 10 to 20 m long, 7 to 15 m wide, and 12 to 13 m deep. The ponds may be either outside the reactor containment vessel, as they are in American reactor designs, or inside, as in some German designs. They contain chemically treated water to remove decay heat, prevent corrosion of fuel elements, and prevent the release of radioactivity from failed fuel cladding. After three to six months, short-lived nuclear products dissipate their energy, and the fuel assemblies can be removed from the ponds and air cooled. However, they are normally kept under water until processed in reprocessing plants. Plans call fuel for reprocessing within a year from removal from the reactor. However, delays in reprocessing throughout the world have resulted in an increase of spent fuel stored at the power plants or in central repositories. In many cases, the ponds are approaching their design capacity. To compensate, the rods are stored more densely until additional repositores are built.[34]

In the closed fuel cycle, spent fuel is sent to reprocessing plants where it is again placed in cooling ponds. From this storage, fractions are taken to a process building; the structure at Barnwell, South Carolina (yet to be opened), is illustrative; it is 175 ft. long by 60 ft. wide by 60 ft. high and divided into cells with walls 3 by 5 1/2 ft. thick. Fuel rods are placed in the cells and mechanically cut into inch-long pieces and then dropped into an acid dissolver that attacks the fuel but not the metal cladding, which is removed and placed in concrete silos. An organic solvent extracts plutonium and uranium from the liquid fuel mixture. Plutonium is separated chemically from uranium and through further processing is converted into a powder to be stored in steel canisters.[35]

From the reprocessing plants, plutonium oxide powder (PuO_2) and uranium dioxide (UO_2) are transported to mixed-oxide fuel fabrication facilities to be converted into fuel pellets. An ideal plant would produce about 300 MT per year, or enough fuel for twenty-five reactors. Plutonium is the principal concern.[36] It would be located in storage areas in vessels containing 250 to 300 kg each, MO_x fuel blenders containing 10 to 20 kg in shielded processing cells (glove boxes), as dust in the air of these cells, and in the mixed-oxide fuel pellets.[37]

The fuel cycle creates a number of wastes. The most voluminous and toxic are high-level liquid wastes resulting from reprocessing. Each ton of reprocessed fuel will produce 330 gal. of long-lived radionuclides that must be permanently isolated from the environment. Although the ultimate disposition of waste product is the subject of considerable study, no final solution has been found. Most scenarios point to solidification into a glass (vitrification) for disposal deep underground in such stable geologic formations as salt or granite. Currently waste products are stored in steel tanks, some with double containment, capable of holding from 100,000 to 1 million gal. The tanks themselves are located in concrete vaults 10 ft. underground.[38] To remove decay heat and prevent the material from melting, water is circulated through several closed-loop coils cooled by a heat exchanger. Spare heat exchanger cooling loops provide a backup in case of accident, as do duplicate diesel-driven pumps for recirculation. Figure 2-12 depicts a model storage in facility.

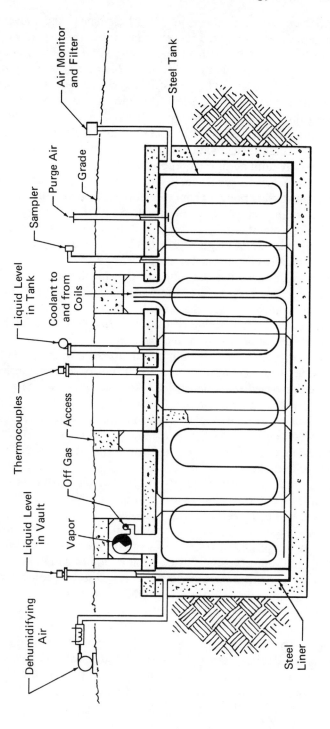

Figure 2-12. High-Level Liquid Waste Storage Tank

Source: U.S. Atomic Energy Commission, *The Safety of Nuclear Power Reactors and Related Facilities*, WASH 1250 (Washington, D.C.: Atomic Energy Commission, July 1973), p. 4-77.

Sabotage scenarios suggest that if explosives are introduced into any of the areas where nuclear material is located, radionuclides can be released in the form of aerosols through adversary created apertures. The heat exchangers for spent fuel and high-level liquid waste are additionally vulnerable. If coolant is cut off, water boils away in from days to weeks, and gaseous material is released into the atmosphere as the material melts.[39] In the case of spent fuel, some analysts believe—a claim disputed by others—that a small nuclear detonation is possible if an explosive jams fuel elements together.[40]

Consequence Calculations from Damaged
Support Facilities

Consequence models of nuclear support facility releases have not been worked out totally. One measure is Science Application Inc.'s ranking of events. The events estimated to produce one or more early fatalities per occurrence are, in descending order of severity:

1. Nuclear device explosion.
2. PuO_2 dispersal in building ventilation.
3. High-level liquid waste tank sabotage at reprocessing plant.
4. Reactor sabotage leading to PWR 1 release.[41]

The events estimated to produce one or more late fatalities per occurrence are, in descending order of severity:

1. Nuclear device explosion.
2. High-level liquid waste tank sabotage at reprocessing plant.
3. Reactor sabotage leading to PWR 1 release.
4. PuO_2 dispersal in building ventilation system.
5. PuO_2 dispersal by explosive loading.
6. PuO_2 dispersal by fire lofting and aircraft release.
7. Stolen spent fuel dispersal by fire lofting.
8. PuO_2 storage sabotage at fuel fabrication plant.
9. $Pu(NO_3)_4$ storage sabotage at reprocessing plant.
10. Stolen high-level waste dispersal by fire lofting.
11. PuO_2 conversion facility sabotage at reprocessing plant.
12. Reactor sabotage leading to PWR 7 release.
13. MO_x fuel blender sabotage at fuel fabrication plant.
14. High-level liquid waste concentrator sabotage at reprocessing plant.
15. Dissolver solution sabotage at reprocessing plant.[42]

These do not apply to destruction by nuclear weapons. The tables provide some basis for comparing where PWR 1 and 7 releases stand in relation to those from other components of the fuel cycle. However, because the rankings are not explicitly weighted, they provide no definitive judgment about the release magnitudes. Note that the late fatality list does not include the relative standing of spent fuel releases either at the reactor or reprocessing plant.

Beyea's study of the planned Gorleben waste treatment facility in lower Saxony, West Germany, provides additional information concerning high-level liquid waste and spent fuel.[43] The Gorleben plant—the future of which is uncertain—is expected to house five 1,400 MT high-level waste storage tanks and a spent fuel pond capable of containing sixty reactor cores. Its cesium-137, ruthenium-106, strontium-90, and plutonium-239 inventories will be the principal biological concern. Beyea accounts for the first two. Assuming a loss of coolant that could not be repaired due to on-site contamination or unstable social conditions (likely during wartime), up to 90 percent of the cesium waste tank inventory involving 1 to 1.4×10^8 Ci of activity—could be released per tank. Contamination could extend 1,500 km to 2,300 km in stable (condition D) weather—the higher figure reflecting a slower deposition rate—contaminating 237,000 km² to 410,000 km² above the 10 rem in thirty year threshold considered safe for habitation. An even more serious release could result from a loss of coolant and hydroden-zircaloy cladding reaction in the planned compact spent fuel pool, which due to the density of materials and age of products—less than one to two years old—requires liquid coolant. Figure 2-13 depicts the worst release across Europe to the point where rain can be expected to wash out material, thus preventing more distant contamination (indicated by the dashed lines). Of course, lesser releases are possible. A spectrum is presented in table 2-20, but it does not indicate the probability of the releases.

With an approximately one-year half-life, the ruthenium threat principally derives from the spent fuel rather than the older wastes. Absorbed through inhalation from the passing plume, this product can induce early (within one year) and late cancers, the latter at a rate of 1 to 10 per 1,000 exposed to 100 rem. Figure 2-14 depicts the worst possible release, and table 2-21 indicates gradations. Although a release of this product will be less extensive than cesium, still thousands of km² would be involved.

The hazards posed by spent fuel and high-level wastes at reprocessing plants depend upon the exposed population. Although extensive, most of the contamination is of relatively low intensity, threatening an increase of a few tenths of a percent in the cancer rate. However given the tens of millions of people who would be vulnerable in the European context, the absolute casualties could run into the tens of thousands.

Figure 2-15 provides additional information about the consequences

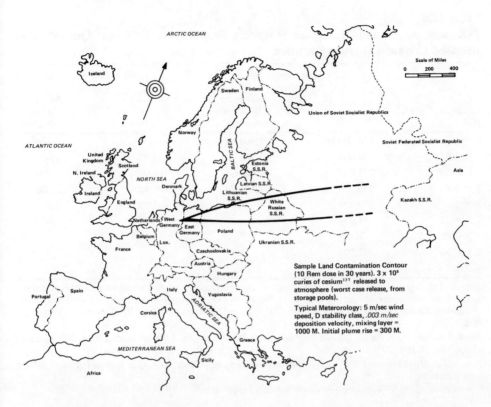

Source: Jan Beyea, "The Effects of Releases to the Atmosphere of Radioactivity from Hypothetical Large-Scale Accidents at the Proposed Gorleben Waste Treatment Facility," unpublished paper, (Princeton: Center for Environmental Studies, Princeton University, 1979), p. 7. Reprinted with permission.

Figure 2-13. Gorleben Reprocessing Plant: Worst Spent Fuel Cesium Release Contour

from nuclear weapons releases. The graph compares the gamma dose rate in time for a Mton fission weapon, a 1,200 MW(e) reactor core, ten-year storage at the reactor, thirty-day high-level storage, and ten-year high-level storage at the reprocessing plant. The considerably greater long-lived contribution of all elements of the fuel cycle compared to the nuclear weapon is notable, with the greatest prominence estimated for the high-level liquid wastes. In the American context, the Chesters assert that 850,000 MW(e) of nuclear generation coupled to an equivalent of 1,770 LWR cores stored in nuclear fuel reprocessing plants or temporary (ten year) high-level waste storage if added to nuclear weapon fallout increase residual radioactivity after one year by an amount equivalent to 30,000 Mtons of 50 percent fis-

Table 2-20
Estimated Areas and Maximum Distances Reached for Different Quantities of Released Cesium-137 from Gorleben

Cesium-137 (Ci)	0.1 m/sec. Deposition Velocity		0.003 m/sec. Deposition Velocity	
	Area	Maximum Distance Reached	Area	Maximum Distance Reached
4×10^8	430,000 km^2	1,900 km	740,000 km^2	2,400[a]
3×10^8	370,000	1,800	680,000	2,400[a]
1.4×10^8	237,000[b]	1,500	410,000	2,300
1×10^8	190,000[b]	1,400	290,000	2,000
4×10^7	100,000		100,000	1,100
1.2×10^7	34,000		17,000	470
3×10^6	7,100			
1.2×10^6	2,300			
4×10^5	550			

Source: Jan Beyea, "The Effects of Releases to the Atmosphere of Radioactivity from Hypothetical Large-Scale Accidents at the Proposed Gorleben Waste Treatment Facility," unpublished paper (Princeton: Center for Environmental Studies, Princeton University, 1979), p. 10. Reprinted with permission.

Note: 5 m/sec. wind speed, ground shielding = 0.25, D stability, 1,000 m mixing level, 300 m initial plume rise, 10 rem/30-year threshold.

[a]Cut off at 2,400 km, assuming rain occurs and washes out radioactivity, thus preventing more distant contamination.

[b]Assumes range of 90% cesium release per waste storage tank.

sion weapons. This figure could be 60,000 Mton by the year 2020. They conclude, "Therefore, the reactor cores and nuclear waste storage facilities can make a very respectable contribution to the fallout problem, especially if the situation one year or longer after the attack is considered."[44]

Military Acts Capable of Exploiting
Nuclear Facility Vulnerability

Given the vulnerabilities of nuclear facilities, how can they be exploited by military acts, including nuclear and conventional weapons bombardment and military sabotage? The effectiveness or lethality of a munition used to damage or destroy a nuclear plant is a function of accurate delivery of sufficient energy (yield). Although increases in either accuracy or yield improve lethality, it rises much more rapidly with improvements in accuracy because lethality

is directly proportional to two-thirds power of the yield, and inversely proportional to the square of the CEP (circular error of probability) . . . [thus]

Source: Beyea, "The Effects of Releases to the Atmosphere of Radioactivity," p. 15. Reprinted with permission.

Figure 2-14. Gorleben Reprocessing Plant Worst Spent Fuel Ruthenium 106 Release Contour

multiplying the yield by a factor of eight only increases the lethality by a factor of four, while reducing the CEP by a factor eight will increase the lethality by a factor of sixty-four.[45]

Studies of the vulnerability of facilities to nuclear weapons bombardment underscore this point. Although nuclear weapons always had sufficient energy to destroy atomic installations, accuracy on the order of one-half mile or greater during the 1960s and early 1970s contributed to the conclusion that the price of a successful attack in terms of the numbers of required strategic missiles was prohibitive. Seventy-eight 100-kton, eighteen 1-Mton, or two to three 10-Mton weapons were believed necessary for a 50 percent probability of success.[46] However, in 1976 these estimates were revised in light of projected ballistic missile delivery accuracies of 30 m or less in the 1980s. With such accuracies, it was concluded that nuclear installations would become increasingly attractive targets. Estimates of the ef-

Table 2-21
Areas and Maximum Distances Reached for Different Quantities of Released Ruthenium-106 from Gorleben

Ru-106 (Ci)	0.1 m/sec Deposition Velocity		0.003 m/sec Deposition Velocity	
	Area	Maximum Distance Reached	Area	Maximum Distance Reached
5×10^8	77,000 km^2	900 km	280,000 km^2	1,900 km
1.7×10^8	26,000	550	77,000	1,000
5×10^7	6,700	270	12,000	390
1.7×10^7	1,800	140	2,500	170
5×10^6	250	67	280	72

Source: Beyea, "The Effects of Releases to the Atmosphere of Radioactivity," p. 17.
Note: 5 m/sec wind speed, D stability, 1,000 m mixing layer, 300 m initial plume rise, breathing rate = 2.7×10^{-4} m^3/sec, 3.9×10^6 rem lung dose received in ten years per Ci inhaled.

fectiveness of a 1 Mton air-burst weapon against a reactor at varying distances suggest that at 8 km, transmission lines carrying external power will fail. The overpressure may also damage intake air filters for auxiliary diesel generators, so restricting air flow that they could not operate, resulting in a meltdown and loss of containment within several hours. At 3.7 km, the explosion would damage the control room auxiliary equipment transformers to the point that the core could not be prevented from melting. At 2.5 km, damage to the containment vessel would impair systems designed to use ice for steam suppression. The primary cooling loop might suffer some minor damage. In this event, primary coolant would boil away and release products in about four hours. At 800 m, weapons overpressure would breach the containment vessel and break the pipes that carry coolant, resulting in depressurization of the pressure vessel in seconds and a product release within minutes. At approximately 600 to 700 m, the reactor's products would be entrained in the weapon's cloud.[47]

In the open literature, the attention to nuclear weapons destruction is not replicated with regard to conventional bombardment. Nonetheless on the basis of the ability of conventional munitions to destroy concrete and steel reinforced concrete targets, it is possible to extrapolate their effectiveness against nuclear facilities. The evidence suggests this capability is increasing rapidly. Bombs, rockets, and artillery are not only becoming more powerful but, more significant, increasingly accurate.

The energy release of these munitions is a function of their composition and the shaping of the charge to perform specific tasks. A 100-lb. general-purpose bomb can penetrate more than 2 ft. of concrete and 4 in. of steel. Since its power is proportional to its size, its 2,000 lb. counterpart can pierce

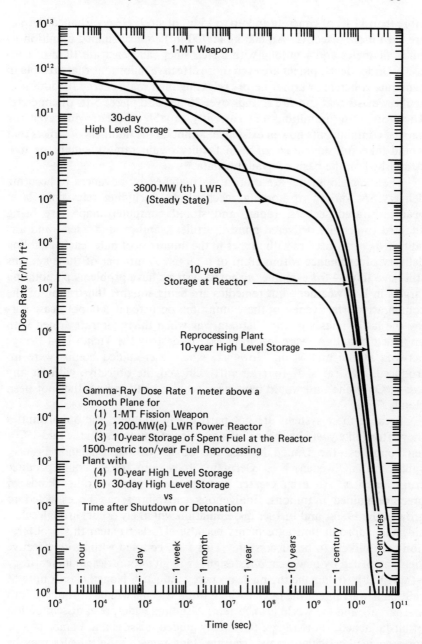

Source: Conrad V. Chester, "Civil Defense Implications of the U.S. Nuclear Power Industry During a Large Nuclear War in the Year 2000," *Nuclear Technology* 31 (December 1976):333.

Figure 2-15. Nuclear Weapon and Nuclear Facility Gamma-Ray Dose Rate versus Time after Detonation and Shutdown

more than 11 ft. of concrete and up to 15 in. of steel. Heavy, shaped charges are even more effective. An 800 kg (1,700 lb.) conical-shaped munition 89 cm in diameter and 1 m long with a steel liner can penetrate 10 m of concrete. Under development are even more effective munitions that are able to penetrate concrete or armor before releasing their main energy. In addition, artillery exists that can fire rounds over 20 mi. and pierce 5 ft. of concrete. Therefore some munitions are currently capable of destroying even the hardest containments now in existence, although the numbers of strikes that would take full advantage of each facility's vulnerability cannot be well established on the basis of available data.[48]

These developments are more than matched by advances in accurate delivery. A variety of homing mechanisms, including television, laser beams, infared seekers, radar, and stored computer maps, are being adapted to aerially delivered bombs, artillery, and ground-, sea-, and air-launched cruise missiles with ranges in the hundreds of miles each, enabling delivery of ordinance within 10 m of a target. A number of these systems still have flaws—for example, airborne systems have problems pinpointing targets in bad weather—but remedies are being sought. Illustration of the comparative effectiveness of these munitions occurred in a famous case during the latter phases of the Vietnam war when thirty aircraft armed with unguided munitions were lost in efforts to destroy the Thanh Hoa Bridge between Hanoi and Vinh. However, when laser-guided bombs were introduced, two raids of four aircraft achieved the objective without any losses. Only one raid would have been required had the weather not been cloudy.[49]

Because these systems are not particularly complex, the opportunities they afford likely will be shared by a number of countries. In the mid-1970s ten countries—the United States, the Soviet Union, England, France, Japan, India, Sweden, West Germany, Italy, and Israel—could produce cruise missiles. An even greater number could produce other kinds of precision-guided munitions. Indigenous capabilities can be expected to grow in the 1980s, and further dissemination will likely occur through sales, aid, and coproduction agreements because efforts to stem this proliferation are unlikely to be successful.[50] This fact, coupled with more effective yields afforded by new weapons designs, will give countries an increasingly lethal ability to damage the nuclear energy industry. Nonetheless a note of caution is warranted: the rate at which these systems will be introduced into arsenals around the world is uncertain. Although widespread dissemination of air-launched, short-range, terminally guided missiles and bombs can be foreseen in the 1980s, highly accurate, long-range cruise missiles may be another matter. They may first appear in the American arsenal in the mid-1980s. However, other countries will be limited in their ability to produce them because they lack accurate information about the flight path to the target. In the American case, satellite mapping techniques will provide on-board computers with information about the terrain that missiles must follow to reach their target. Still even without such sophisticated modes of delivery, more traditional conventional bombardment can release radionuclides.

Nuclear facilities are also vulnerable to military sabotage, despite the fact that many installations incorporate security measures designed to impede or prevent access to facilities. These measures vary. At some facilities, security is limited to fences topped with barbed wire. Others maintain special alarms and infrared and closed-circuit television systems, but attack-resistant guard houses are often manned by poorly trained guards. In France, the Electricité de France built an artificial hill beside its Super Phoenix fast breeder reactor to discourage attacks by terrorists.

Although these measures provide protection against lightly armed, rather unsophisticated intruders, most studies suggest that facilities are vulnerable to destruction by anyone with technical competence, including adequately trained military units assuming they penetrate plant external defenses. According to testimony by a former American military demolitions expert,

> I could pick three to five ex-Underwater Demolition Marine Reconnaissance or Green Beret men at random and sabotage virtually any nuclear reactor in the country. It would not be essential for more than one of these men to have had such experience.
>
> Access for purposes of taking over and placing charges could be gained by force under ruse. Alternatively, containment could be breached from the outside with relatively small shaped charges and additional charges could be quickly set after gaining entry through the breech. The "engineered safeguards" would be minimally effective or wholly ineffective and the amount of radioactivity released could be of catastrophic proportions.[51]

This testimony is consistent with the Ford-Mitre conclusions. According to *Nuclear Power Issues and Choices*:

> While it is true that safety features reduce the likelihood of a major incident, they cannot reduce it to an inconsequential level. In contrast with an accident where "defense-in-depth" deals with chance coincidence of malfunction, probabilities here must take into account deliberate simultaneous sabotage or reinforcing safety measures.
>
> It is also true that it would require technically sophisticated and knowledgeable commandos to have a high probability of causing a large radioactive release. However, this does not pose an insuperable barrier to a group with time, resources, and determination. The flow of personnel through military nuclear programs and the growing international civilian nuclear industry provide a large pool of experienced manpower from which

a group could seek assistance. Reactor personnel held as hostages might be forced to assist their captors under duress. The technical problems in blowing up a reactor would be easier than those in designing and constructing a nuclear explosive. Explosives could be carried by a few people into a reactor or other facility and could cause major damage. Shaped charges could severely damage main inlet pipes for cooling water. Automatic control and safety equipment could be destroyed. Even primary containment could be ruptured with conventional explosives.[52]

The consequences of sabotage-induced damage can be gleaned from tables 2-4 to 2-7. It is reasonable to conclude that such scenarios could apply to support facilities as well as reactors.[53]

Conclusions

Nuclear energy facilities are vulnerable to destruction in time of war. However, destruction would not be easy. Facilities usually are housed in massive reinforced concrete structures, built to rigorous standards, and have a number of backup systems to compensate for primary system failure and to minimize the consequences of accidents. The exact standards vary depending upon the manufacturer; German facilities often have the largest number of compensatory systems and the Soviet Union the least. But given current military technology, they are not sufficient to prevent the release of nuclear products into the environment as the result of a nuclear weapons attack, a concerted conventional weapons bombardment, or efforts by sophisticated saboteurs. The problem is minimized today because nuclear weapons are limited to a few countries, and precision-guided munitions capable of destroying facilities still are not widespread. However, this situation will change as increasingly lethal conventional munitions are introduced into the arsenals of many countries.

In the event of radionuclide release into the atmosphere, the consequences depend upon a number of variables. They would be maximized if measures to filter products such as sprays and ice do not function; materials containing large fractions of actinides and fission products vent in stable weather through the top rather than the bottom of a structure and deposit at a high rate. High-level liquid wastes, spent fuel at reprocessing plants and reactors, pose the most serious hazard. Reactors in particular would be attractive targets given their relatively large number, inherent value, contribution to the economy, vulnerability, and concentration of radioactive products. Major reactor accident consequences models suggest that contamination resulting in deaths within sixty days could extend forty miles downwind in very stable weather. The zone of greatest concern usually would be 0 to 15 mi. from the release point. Low-level contamination, which would be absorbed over time and result in cancers and genetic effects, and radio iodine, which has a short half-life but which can be absorbed quickly, would cover considerably greater areas. It is estimated that a large release—BWR 2—from a 580 MW(e) reactor, which is 60 percent of the size of many power plants in operation, could contaminate on the average 10,000 km^2 with radiation beyond what the *Reactor Safety Study* defined as safe. The human effects from any such contamination are minimized through relocation in radiation-proof shelters or unexposed regions, prompt medical treatment for the irradiated, and impounding of contaminated food. Such measures would not reduce ground contamination. Long-term or permanent relocation of inhabitants or decontamination would be necessary but effective only to a point. If such measures were not taken expeditiously and conscientiously, the incidence of effects would rise accordingly. In densely populated regions, thousands would succumb to early, late, and genetic consequences. This danger led Beyea at one point in the Barsebäck study to suggest that in the case of an accident, "evacuation plans for Copenhagen and Malmö and other towns no longer seem unreasonable".[54] This conclusion appears reasonable in view of the evacuation plans that were contemplated in the event of a meltdown during the 1979 Three Mile Island reactor accident in Pennsylvania. It would be applicable also to releases resulting from military acts. Furthermore in agricultural regions, even if decontamination proved successful, consumers might still fear tainting of produce, making marketability difficult or impossible. Finally the experience of the Japanese atom bomb survivors suggests that persons irradiated, however slightly, often carry psychological scars throughout their lives and transmit their fears to their children.

3 Strategic Implications

Because nuclear energy facilities contain such large inventories of biologically threatening radionuclides, they can make potentially useful radiological weapons when manipulated for strategic purposes. Of the authors touching upon this danger, Chester Cooper, assistant director of the Oak Ridge Laboratory's Institute for Energy Analysis, explores the possibilities most thoroughly.[1] Cooper argues that nuclear power plants, which have been discounted in the strategic trade-offs between the North Atlantic Treaty Organization (NATO) and the Warsaw Pact, could have significant implications for the third world where plants would be hostages to their neighbors. This could have the positive effect of constraining those governments' own bellicose behavior and serve as levelers of military power between weak and strong neighbors. Summing up his argument Cooper writes:

> What can be said for the concept of nuclear power plants as potential hostages is that by installing a reactor on its territory, a country increases its vulnerability to grave, possibly unacceptable damage in the event of war. As a result, that nation's leaders might be inclined to raise the threshold of their own inclinations toward aggression. Admittedly this is a frail substitute for robust international agreements, but in the present order of things it is not a trivial consideration. . . .

> The idea must not be pressed too far. The export of a nuclear power plant to a Third World country cannot be advocated simply as a means to constrain its own military adventurism. It would be best to confine nuclear exports to stable, responsible countries at peace with their neighbors—countries that would also be less likely to divert nuclear materials. But, alas, few nations anywhere can be counted on to meet such exacting standards indefinitely. In any case, other nuclear exporters might not accept American standards and would gladly fill the orders of would-be customers.

> Aside from the promise of a vast increase in energy supply for developing nations, nuclear powered generating stations could actually improve relations among countries. The risk of widespread radioactive contamination by nuclear power plants hit by even conventional bombs could introduce a positive new element into the military calculations of powers outside the NATO-Warsaw Pact arena. As they balance military and diplomatic solutions to local conflicts, moderation rather than bellicosity might become the better part of valor.[2]

Cooper's proposition suggests that nuclear power plants, as well as other large reservoirs of radionuclides, can be manipulated in deterrent strategy, coercive diplomacy, and military strategy. Deterrent strategy threatens unacceptable costs to persuade an opponent not to initiate an action. Coercive diplomacy is a politico-military effort that uses the threat of force supplemented by controlled selective military violence coupled to the threat of greater violence to persuade an antagonist to stop short of his goals or to undo an action. Military strategy is decisive use of force against an adversary's military capabilities to alter his will.[3]

Nuclear weapons provide an obvious parallel to nuclear facility radioactivity perceived as a weapon along all three of these lines. A case in point, the United States uses its strategic missile and bomber forces to threaten the Soviet Union with what Thomas Schelling describes as "monstrous damage . . . without first requiring the achievement of victory," the message being that the costs of war far outweigh the benefits.[4] Theater tactical nuclear weapons supplement forces in Europe. American defense planners believe that these weapons may reinforce deterrence and coercive diplomacy through a military capability to destroy Soviet armed forces.[5]

Nuclear facility radioactivity affords combatants manipulative opportunities both similar and distinct from those of nuclear weapons. Like nuclear weapons, large releases from facilities can contaminate extensive property. In one sense this contamination is more pernicious since it is long-lived. Casualties could run in the tens of thousands. From the psychological point of view, traumatization may be as great as that suffered by Japanese survivors of atomic bombings. Such facts appear to have encouraged Cooper to assert that a core meltdown in a nuclear reactor might have the same effect on the world as mutual assured nuclear weapons destruction has had in constraining aggression between NATO and the Warsaw Pact.

There are profound differences as well. What makes nuclear weapons unique is not so much the number of people they can kill—during World War II millions died in a conventional conflict—or perhaps not even the devastation they bring but rather the yield to weight efficiency, speed, and reliability with which this devastation can be achieved. As Schelling puts it, "Something like the same destruction always *could* be done. With nuclear weapons there is an expectation that it *would* be done."[6] By contrast, nuclear facility radiation is subject to numerous variables that diminish its reliability. Furthermore even in the worst release, the radiation is not as effective as that produced by nuclear weapons, particularly in inducing early fatalities. The prompt doses that facilities deliver simply are not large enough. Furthermore because nuclear energy installations are stationary, they cannot follow evacuees. Nuclear weapons can follow, thus ensuring a large number of casualties unless populations have access to blast- and radiation-resistant shelters.

Having made these distinctions, for what purposes can facility radiation be used? As an offensive countercombatant weapon, radiation has limited utility because it is unreliable and the facility is not portable. Only the coincidental location of facilities in the proximity of military installations, troop concentrations, or perhaps strategic geography, such as valleys, would allow for such use. Even so troops could avoid the effects of radiation by relocating, although war planning would be further complicated. Doses as low as 80 to 120 R will induce radiation sickness—nausea, vomiting, and/or diarrhea for about a day—in 5 to 10 percent of the exposed and manifestation of these symptoms for 25 percent after a two- to three-week latency period.[7] A fraction of those exposed may suffer functional impairment for several months. The resulting diminished morale of the exposed and nonexposed alike would limit combat effectiveness and increase the burden on medical facilities. These problems would be exacerbated at higher exposures.

Military nuclear facilities might serve more effectively as a component of defense strategy. In this case, the installations, particularly waste storage, could be situated along invasion routes. The defender could threaten to release or actually release the products (as the Chinese and Dutch destroyed their dikes), thereby raising the attacker's risks, albeit at the cost of self-contamination. The prospective invader would need a means of circumvention, thus complicating his strategy and perhaps inhibiting certain acts.

Such facilities might have even greater utility as a component of coercive diplomacy, particularly in times of crisis, or limited war. Assuming that the vulnerability of the aggressor is less than that of the target party either because of the numbers of installations, their hardness, weather, the effective munitions that can be delivered, and/or their proximity to populations or agriculture, threats of destruction could be used to intimidate and compel policy changes.

Perhaps the most significant use of facility radioactivity lies in deterrence similar to the way the United States uses its nuclear weapons to deter the Soviet Union. States could declare that military acts against them would be met by destruction of the aggressor's facilities. Deterrence would succeed, assuming that the potential adversary believes the threat can be carried out and is psychologically sensitive to the outcome. Population protection might still be possible through successful evacuation and sheltering, but relocation before hostilities broke out could provoke and invite preemptive attack. However, long-lived facility radioactivity would be unavoidable and exacerbated if many installations were damaged or destroyed.

This policy could be reinforced where belligerents border one another. Here defenders anticipating a long-term threat could build facilities along frontiers where prevailing surface winds blow toward the antagonist. Doing so, of course, would be at the risk of increasing one's own vulnerability to destruction and self-contamination given unpredictable winds and associated economic and social costs. Still this strategy might be worth the gamble if it was felt that the relative risks for the antagonist were greater. In any case, this option must be mentioned for, as Yehezkel Dror, who has studied unconventional state behavior, has noted, "When trying to deal with potentially catastrophic future possibilities, we must be able to envisage unprecedented seemingly remote occurrences which have not formed strong inprint on our frame of reference."[8]

Used in the deterrence modes described, facility radioactivity has certain advantages over nuclear weapons in terms of stable deterrence. Because the installations are not portable and the effects of radiation contamination in most instances would not be immediate, facilities, unlike nuclear weapons, cannot be used as major first-strike countercombatant weapons. Still because of their contamination potential, they could be considered more unambiguously a defensive or inoffensive nuclear deterrent.[9] This factor could diminish pressures that some predict will mount in the future for the acquisition of offensive nuclear weapons by nonnuclear weapons states and thereby increase strategic stability.

The discussion below examines this proposition, as well as the utility of nuclear installations, from the military and coercive diplomatic points of view in the empirical contexts of the Soviet Union, Western Europe, the Middle East, Korea, South Asia, West Asia, South Africa, and the United States.

Soviet Union

This survey of nuclear facility vulnerability begins with the Soviet Union because second only to Israel, which at the present time does not operate any large nuclear installations, it confronts more potential antagonists than any other nation. A review of the Soviet nuclear program provides some perspective. The Soviets have either in operation or under construction over forty reactors producing from 100 Mw(e) to 1,000 Mw(e). Figure 3-1 displays their approximate location as well as population distribution. The map key (table 3-1) provides more detailed information about the size and type of each reactor, population distribution including the nearest urban center, prevailing geostrophic winds over surrounding land, and precipitation probabilities. In addition to these installations, the Russians operate at least one commercial enrichment plant and a pilot reprocessing plant, as well as an undetermined number of such installations for its nuclear weapons program. A reprocessing facility is planned for a commercial breeder reactor program that will go on-line in the mid-1980s.[10]

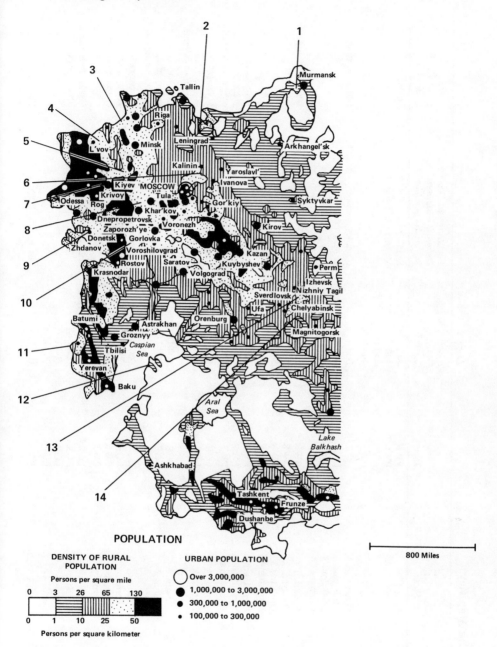

POPULATION

DENSITY OF RURAL
POPULATION

Persons per square mile

0   3   26   65   130

0   1   10   25   50

Persons per square kilometer

URBAN POPULATION

○ Over 3,000,000

● 1,000,000 to 3,000,000

● 300,000 to 1,000,000

• 100,000 to 300,000

800 Miles

Source: Central Intelligence Agency, "USSR Summary Map-Population," 501614 (Langley, Va.: Central Intelligence Agency, April 1974).

Figure 3-1. Approximate Location of Soviet Power Reactors

Table 3-1
Soviet Union Map Key

Nuclear Plant No.	Designation of Plant	Type[a]	Power MW(e) (100+)[b]	Location and Nearest Urban Populations within Forty Miles[c]	Land Use in Vicinity
1	Kola 1	PWR	440	Kola Murmansk 396,000	Forest
	Kola 2	PWR	440		
	Kola 3	PWR	440 U/C		
	Kola 4	PWR	440 U/C		
2	Leningrad 1	LGR	1,000	60 miles west of Leningrad	Dairy farming
	Leningrad 2	LGR	1,000		
	Leningrad 3	LGR	1,000		
	Leningrad 4	LGR	1,000		
3	Drukshai 1	BWR	1,500 U/C	Vilnus 514,000	Dairy farming
	Drukshai 2	BWR	1,500 ?[d]		
	Drukshai 3	BWR	1,500 ?		
	Drukshai 4	BWR	1,500 ?		
4	West Ukraine 1	PWR	440 U/C	Rovno 147,000	Livestock, grain, other crops, woodland
	West Ukraine 2	PWR	440 U/C		
	West Ukraine 3	PWR	1,000 U/C		
5	Smolensk 1	LGR	1,000 U/C	Smolensk 278,000	Livestock, grain, other crops, woodland
	Smolensk 1	LGR	1,000 U/C		
6	Kalinin 1	LGR	1,000 U/C	Kalinin 429,000	Livestock, grain, other crops, woodland
	Kalinin 2	LGR	1,000 U/C		
7	Chernobyl 1	LGR	1,000	Chernobyl Kiev 2,160,000	Livestock, grain, other crops, woodland
	Chernobyl 2	LGR	1,000		
	Chernobyl 3	PWR	1,000 U/C		
	Chernobyl 4	PWR	1,000 U/C		
8	Kursk 1	LGR	1,000	Kursk 363,000	Grain, livestock, other crops
	Kursk 2	LGR	1,000		
	Kursk 3	LGR	1,000 U/C		
	Kursk 4	LGR	1,000 U/C		
9	South Ukraine 1	PWR	1,000 U/C	Nikolaev 428,000	Livestock, grain, other crops

10	Novovoronezh 1	PWR	210	Novovoronezh	Livestock,
	Novovoronezh 2	PWR	210	Voronezh 880,000	grain, other crops
	Novovoronezh 3	PWR	440		
	Novovoronezh 4	PWR	440		
	Novovoronezh 5	PWR	1,000		
11	Armenia 1	PWR	440	Oktemberyan	Cotton, fruit,
	Armenia 2			Yerevan 1,055,000	vineyards
12	BN 350	LMFBR	150	Shevchenko	Desert
13	Beloyarsk 1	LGR	100	Sverdlovsk	Forest
	Beloyarsk 2	LGR	200	1,200,000	
	Beloyarsk 3	LMFBR	600 U/C		
14	Siberian 1	LGR	100	Troitsk 76,000	Grain,
	Siberian 2	LGR	100		forest
	Siberian 3	LGR	100		
	Siberian 4	LGR	100		
	Siberian 5	LGR	100		
	Siberian 6	LGR	100		

Sources: Joseph Lewin, "The Russian Approach to Nuclear Reactor Safety," *Nuclear Safety* 18 (July-August 1977):438-450; "World List of Nuclear Power Plants," *Nuclear News* 22 (August 1979):86-87; Nuclear Engineering International, "Map of the World's Nuclear Power Plants" (London: IPC Business Press Ltd., 1977); Population Division, Department of Economic and Social Affairs, U.N. Secretariat, "Trends and Prospects in the Populations of Urban Agglomerations, 1950-2000, as Assessed in 1973-1975," ESA/P/WP.58 (New York: United Nations, November 21, 1975), pp. 57-60; Editors of Life and Rand McNally, *Life Pictoral Atlas of the World: Comprehensive Edition* (New York: Time Inc., 1961), p. 267; Central Intelligence Agency, "USSR Summary Map—Land Use," 501614 (Langley, Va.: Central Intelligence Agency, April 1974); J.G. Bartholomew et al., *Atlas of Meteorology* (Chicago: Denoyer-Geppert Co., 1899), plates 12, 20; Paul E. Lydolph, *Climates of the Soviet Union* (Amsterdam: Elsevier Scientific Publishing Co., 1977), p. 354.

Notes: Geostrophic winds over European Russia, where most reactors are situated, tend to be southwesterly in winter and northwesterly in summer. Surface winds are variable. Precipitation is greatest from June through August, averaging two to four inches a month along a latitudinal zone 40 degrees by 60 degrees extending into Siberia to 140 degrees longitude. Most regions get less than an inch of precipitation a month during the remainder of the year.

[a] LGR: light water cooled graphite moderated reactor.

[b] U/C: under construction.

[c] All population data are 1980 estimates except for Troitsk, which is a 1960 estimate.

[d] ?: MW(e) output uncertain.

Factors unique to the Soviet nuclear facilities both enhance and diminish its relative vulnerability and the consequences resulting from destruction. The average population density of 10 to 50 per km² near most reactors is considerably less than that of other European countries. For example, in West Germany the figure ranges from 100 to 300 persons per km². Consequently the number of Russians who could be irradiated beyond the *Reactor Safety Study*'s threshold unless evacuated would average 100,000 to 500,000 using Beyea's Swedish calculations that 10,000 km² would be contaminated assuming a 580 MW(e) reactor undergoing a BWR under stable weather. The contamination would be greater if a larger reactor or several reactors were destroyed and less extensive if the reactor or support facility were smaller or the release less severe.

The Russians further mitigate the threat to their facilities through their active and passive civil defense programs. They maintain the most extensive antiaircraft defenses in the world, supplemented in the vicinity of Moscow by an antiballistic missile system. A passive civil defense program that complements this system requires every citizen to take a twenty-hour civil defense course and be prepared for relocation in time of war. Urban shelters also are provided.[11]

Each of these advantages is offset by countervailing factors, however. Although the Soviet Union is less densely populated than most Western European countries, some of its installations are situated within a few miles of population centers. Soviet power plants are inherently more vulnerable because they often lack containment safeguards and the many redundant emergency systems characteristic of Western reactors. Active defenses may have only limited effectiveness against ballistic missiles, the coming generation of cruise missiles, and some aircraft. And Soviet evacuation and sheltering plans may be only marginally effective. One critique points out that the sufficiency of the Soviet program rests on a number of dubious assumptions: the ability of a country with limited transportation to relocate large populations quickly, sheltering in poorly stocked facilities, and urban shelters relying on air filtration powered by external energy sources. Whatever effectiveness the plan may have in reducing prompt lethal exposure, as currently designed it is of little help against long-term ground contamination considering that the populace is to return to their homes after the attack and await further instructions.[12]

Despite Cooper's assertion that nuclear facilities have no politico-military implications in Europe, at least eight countries in and out of NATO (the United States, Britain, France, Germany, China, Iran, Yugoslavia, and Japan) might choose to take advantage of the vulnerability of Soviet nuclear facilities in order to reduce their own susceptibility to Soviet intimidation and overt military acts. The Russians might be particularly receptive since they are the only people to have suffered a major nuclear ac-

cident.[13] In 1957 what is believed to have been an explosion at a waste
storage facility located in the Urals contaminated hundreds of square miles
and may have resulted in many hundreds of casualties, including fatalities.
Current American strategic targeting doctrine already calls for destruction
of Soviet industry, including electrical generation.[14] Conceivably some
rhetorical focus by the U.S. Department of Defense on Soviet nuclear facility vulnerability to conventional and nuclear munitions and the consequences deriving therefrom—particularly problems for Soviet recovery—could enhance American deterrence power. Facilities could be used
as part of a coercive diplomatic and, if Soviet plants are located near
military bases, military strategy. This policy should prove most attractive
for countries with limited nuclear arsenals that are less certain they can inflict unacceptable damage to deter Soviet coercion. Britain and France
might enhance the credibility of their nuclear deterrent vis-à-vis the Soviets
if they threatened to ground-burst a portion of their nuclear arsenal over
Russian facilities or if they acquired a cruise missile capability that could
penetrate Soviet defenses with lethal conventional warheads. China could
do likewise if it modernized its forces.

West Germany, Iran, Japan, and Yugoslavia are all concerned about
Soviet intimidation. Each has speculated about outside assistance, notably
from the United States, should a crisis develop between them and the Russians. Uncertainty motivates them to contemplate use of nuclear weapons.[15]
However, acquisition of nuclear weapons might reduce rather than increase
security: the Soviets might take measures to prevent such acquisition or
other neighbors might acquire their own, with the end result being a nuclear
arms race.[16] By contrast, the ability to destroy Soviet nuclear facilities might
be less provocative, being void of first-strike implications, and serve principally as a defensive deterrent. The West German Air Force may already
have such a deterrent capacity. German capabilities may increase as they acquire more sophisticated aircraft from the United States. They could be
made more formidable through acquisition of the developing generation of
cruise missiles and intermediate- or medium-range ballistic missiles armed
with conventional warheads. Should Iran renew the military modernization
begun under the shah, it could strike Soviet reactors located along the Cas-
pian Sea and Armenia. In the foreseeable future, Japan and Yugoslavia will
find it difficult to penetrate Soviet defenses. Yugoslavia would have to import the capability over Soviet objections. Japan could forge its own
weapons but would probably need to develop an intercontinental ballistic
missile capable of reaching Soviet facilities in central Asia and Soviet
Europe, an expensive undertaking. If the Russians built installations closer
to Japan, the vulnerability to Japanese attack would increase.

Given objections to other weapons developments directed against them
in the past—most recently the neutron bomb—the Russians may object to a

facility destruction strategy, particularly one adopted by nonnuclear weapons states and perhaps China as well. Destruction of nuclear installations has limited offensive utility, but the threat of it could subtly change the psychological balance of power, limiting maneuverability in fulfilling national objectives. Concern might also arise over the proliferation of delivery systems, notably cruise missiles that could carry nuclear warheads.

At least three responsive courses of action could result. One would be renunciation of the strategy and acquisition of the means to destroy facilities since the increase in tension that might ensue, including threats of counterretaliation, is not worth what may be uncertain benefits. A second alternative would be the threat of acquisition to exact Soviet concessions. For example, West Germany and its NATO allies could use the threat to make the Russians more accommodating in the European Mutual Force Reduction Talks (MFR). The Iranians, Japanese, and Yugoslavs could openly explore the strategy and renounce it contingent upon Soviet noninterference in their internal affairs or to make the Russians generally more accommodating. A last option would be acquisition of a facility destruction capability notwithstanding Soviet objections. States here would feel that the new element introduced into the strategic equation would sufficiently complicate Soviet risk calculations and thereby enhance deterrence to be worth risking Soviet displeasure.

Western Europe

With about 170 nuclear power plants, at least six reprocessing and mixed oxide (plutonium-based) fuel fabrication plants, operating, under construction, ordered, or planned, Western Europe (Austria, Belgium, Denmark, Finland, France, West Germany, Italy, Luxembourg, the Netherlands, Portugal, Spain, Sweden, Switzerland, and the United Kingdom) has the greatest concentration of nuclear energy of any other region in the world outside the United States.[17] However, the facilities may be less attractive targets than are Soviet plants, since Western Europe already faces Soviet ability to inflict massive destruction through nuclear weapons bombardment. Thus there appears to be no rationale for Soviet release of facility radionuclides other than to maximize the difficulty of recovery. But to do so, particularly with ground-burst nuclear weapons, might prove counterproductive since prevailing westernly geostrophic winds could carry radioactivity across Eastern Europe into the Soviet Union itself. Destruction of Western European facilities might prove attractive for other reasons, notably to stop electrical generation, but would not rquire destruction of radionuclide containments assuming accurately delivered munitions that preserve external energy or coolant supplies.

Any release of facility radioactivity, intended or accidental, could have very significant implications for the course of war in Europe. In recent years, the Russians have increased their conventional forces, and some military writings suggest that should war in Europe come, it might be limited to nonnuclear weapons, at least in its early phases.[18] To meet this challenge, NATO has strengthened its conventional forces in hopes that nuclear weapons, with their devastating effects, will not have to be used. The focal point of any such conflict is likely to be West Germany. With at least thirty power plants and other support facilities operating or planned, its wartime vulnerability could have significant implications for the course of conflict. Figure 3-2 show the locations of the facilities and provides supplementary information, including the three principal invasion routes from the east; the Fulda gap; the Hof corridor, and the northern German plain. (See table 3-2.) The northern German plain is believed to be the most attractive because it is flat and easily traversed by tanks. Four large reactors ranging in size from 662 MW(e) to 1,363 MW(e) are located in or planned for this region. Should these facilities be damaged or destroyed by Soviet forces, radionuclides would be introduced into the conflict and could conceivably lower the nuclear weapons threshold. NATO would face a dilemma: respond in kind by attacking nuclear facilities in Eastern Europe and the Soviet Union or use or threaten to use nuclear weapons to stop the Soviets. Either course would escalate the conflict. NATO could also ignore the destruction but would thus encourage further Russian attack.

The West German installations have other implications. Strategically situated along invasion routes, they could function as radiological mines. These may not be insurmountable obstacles for Soviet combat troops protected by tanks and personnel carriers designed to operate in lightly irradiated regions, but unprotected support troops may have a more difficult time. Furthermore intelligence regarding facility locations would allow avoidance of contaminated regions. Yet these facilities could complicate Soviet war planning.

The existence of West German facilities could be incorporated into a coercive diplomacy. West Germany could threaten to or, assuming favorable winds, actually destroy one of its own installations near the East German frontier to manifest its determination to defend itself and to impress upon the Warsaw Pact nations the cost of continued combat. With its own facilities, coupled to an ability to destroy facilities in Russia and Eastern Europe, the West Germans would have some control over NATO's initiation of nuclear war, rather than relying upon the United States, Britain, and France.

Self-destruction could also be applied as an adjunct to a deterrent strategy that includes targeting of Warsaw Pact facilities. Installations placed along the East German border—indeed the Geesthact-Krummel reactor to be completed in 1980 is so situated—may serve to remind East Germans of the consequences should war occur since prevailing westerly geostrophic winds would carry contamination east. For historical reasons the Russians may not be concerned about the fate of any Germans, but the Soviet Union still must rely on the East Germans as a staging area and, as war games suggest, for combat support as well. This reliance gives the East Germans some leverage. Using Beyea's Swedish calculations for a reactor half the size of the Krummel plant, a major release could displace on the average of 750,000 people (the region has an average of 75 persons per km²) over 10,000 km².[19] The magnitude of the consequences would be many times greater if the planned Gorleben spent-fuel or waste-holding tanks were destroyed.

Middle East

Nuclear energy is not now being generated for commercial purposes in the Middle East. The reactors in the region are small research installations, the largest being the Dimona facility generating 26 MW of heat located in Israel's Negev desert. This situation may change in the late 1980s. At least four countries—Israel, Egypt, Libya, and Iraq—plan atomic power plants, although construction has not begun at the time of this writing and uncertainties remain whether implementation will take place.

The experiences of Israel and Egypt indicate problems that have yet to be overcome in the region as a whole. Although both countries were promised atomic power plants by the United States in 1973 as part of the disengagement agreement, consummation has stalled. In Israel's case, this is explained by the American decision to suspend the commitment until Jerusalem signs the nonproliferation treaty. To circumvent the American demand, Israel now is approaching other vendors, including France, Germany, and Canada, and also is considering construction of its own reactor. An added complication is location. Originally the Israelis planned a 960 MW(e) facility at Nitzanim located about twenty miles south of Tel Aviv along the coast. However these plans were aborted in 1979 because the region is seismically unstable, and perhaps security is a concern as well. Consideration is now being given to location in the Negev along the contemplated Dead Sea Canal (see figure 3-3). Some Israelis also have suggested a possible joint venture with Egypt along the Sinai coast. Egypt itself plans to build a plant at Sidi Kreir 30 km west of Alexandria (see figure 3-4 and table 3-3) and contemplates up to nine additional plants by the year 2000. However it has been unable to begin its first facility because of inadequate financing and differences with the United States over safeguards.[20]

As Middle East nations modernize their military forces during the 1980s, they increasingly will be able to destroy nuclear facilities. Major releases could have significant consequences for the region should war

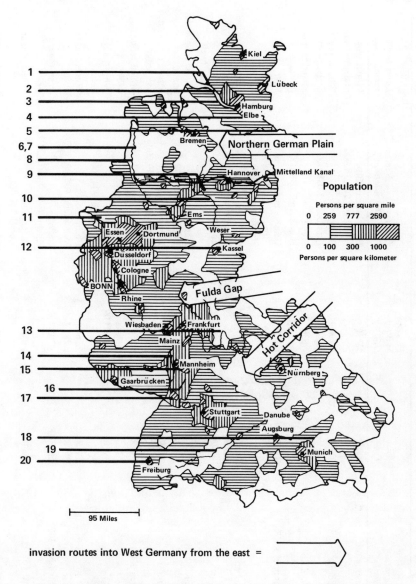

Source: Central Intelligence Agency, "West Germany—Population," 400470 (Langley, Va.: Central Intelligence Agency, May 1972).

Figure 3-2. Approximate Location of West Germany's Nuclear Energy Installations and Invasion Routes

recur. (In 1979 Egypt and Israel signed a peace treaty. While there is hope this will end their antagonism, the region's volatility makes prediction impossible. Thus the argument here presumes the possibility of war between the two, as well as between Israel and its other neighbors, remains.) None of

Table 3-2
West Germany Map Key

Nuclear Plant No.	Designation of Plant	Type	Power MW(e) (100+)[a]	Location and Nearest Urban Populations[c]	Land Use in Vicinity
1	Brunsbuettel	BWR	771	Brunsbuettel/Elbe Hamburg 20,038,000	Wheat, sugar beets, pasture
2	Brokdorf	BWR	1,290 U/C	Brokdorf Hamburg 2,038,000	Wheat, sugar beets, pasture
3	Stade KKS	PWR	630	Stade Hamburg 2,038,000	Mixed cropland, pasturage
4	Kruemmel KKK	BWR	1,260 U/C	Geeshacht-Kruemmel/Elbe Hamburg 2,038,000	Pasturage
5	Gorleben	Waste treatment planned		Gorleben Wittenberge E. Ger. 32,000	Mixed, intensive agriculture, rye, potatoes, wheat, sugar beets
6	THTR 300	THTR[b]	300 U/C	Hamm-Uentrop Rhein-Ruhr 9,949,000	Mixed cropland, pasturage
7	Hamm	PWR	1,300	Hamm-Uentrop Rhein-Ruhr 9,949,000	Mixed cropland, pasturage
8	Lingen KWL	PWR	256	Lingen 22,400	Mixed cropland, pasturage
9	KWG	PWR	1,294 U/C	Grohnde Hannover 904,000	Wheat, sugar beets
10	KKP 1 KKP 2	BWR PWR	864 1,281 U/C	Philippsburg Mannheim 935,000	Wheat, sugar beets
11	Kalkar SNR-300	LMFBR	295	Kalkar Rhein-Ruhr 9,949,000	Pasturage
12	KWW	BWR	640	Wuergassen Kassel 377,000	Wheat, sugar beets
13	Grafenrhenfeld KKG	PWR	1,225	Schweinfurt Wurzburg 199,000	Mixed cropland, pasturage

14	Bilbis A	PWR	1,146	Worms/Rhein Mannheim 935,000	Potatoes, rye barley, vineyards, wheat, sugar beets
	Bilbis B	PWR	1,240		
	Bilbis C	PWR U/C	1,232 U/C		
15	Obrigheim KWO	PWR	328	Obrigheim/Neckar Mannheim 935,000	Forest
16	GKN 1	PWR	805	Neckarwestheim Mannheim 935,000	Forest, wheat, sugar beets
	GKN 2	PWR	805		
17	WAK Karlsruhe	Reprocessing plant		Karlsruhe 503,000	Potatoes, rye, barley
18	Isar KKI	BWR	870	Niederaichbach via Landshut Munich 1,970,000	Potatoes, rye, barley
19	KRB I Block A	BWR	237	Gudremmingen/ Danube	Forest, mixed cropland, pasturage
	KRB II Block B	BWR	1,249		
	KRB III Block C	BWR	1,249		
21	KKU	PWR	1,230	Esenham[d]	
22	Kaerlich	PWR	1,227 U/C	Kaerlich[d]	
23	Neupotz	PWR	1,246 Pl	Neupotz[d]	
24	KWS	PWR	1,284	Wyhl[d]	
25	KWW	BWR	640	Wuergussen[d]	

Sources: "World List of Nuclear Power Plants," p. 72; Nuclear Engineering International, "Map"; of the Congressional Research Service, *Nuclear Proliferation Factbook* (Washington, D.C.: Government Printing Office, 1977), p. 199; Jan Beyea, "The Effects of Releases to the Atmosphere of Radioactivity from Hypothetical Large Scale Accidents at the Proposed Gorleben Waste Treatment Facility," unpublished paper, (Princeton, N.J.: Center for Environmental Studies, Princeton University, 1979); Population Division, "Trends and Prospects," p. 56; Central Intelligence Agency, "West Germany—Land Use," 500470 (Langley, Va.: Central Intelligence Agency, May 1972); Bartholomew et al., *Atlas*, plates 12, 22.

Notes: Geostrophic winds tend to be southwesterly in winter and northwesterly in summer; surface winds are variable. Precipitation averages two to four inches per month much of the year. It is heaviest during the summer and lightest during the winter. The southern half of the country receives more than the northern half.

[a]UC: under construction; Pl: planned.

[b]THR: Thorium high-temperature reactor.

[c]All population data are 1980 estimates, except for Wittenberge and Lingen, which are 1960 estimates.

[d]Insufficient information on location of these facilities.

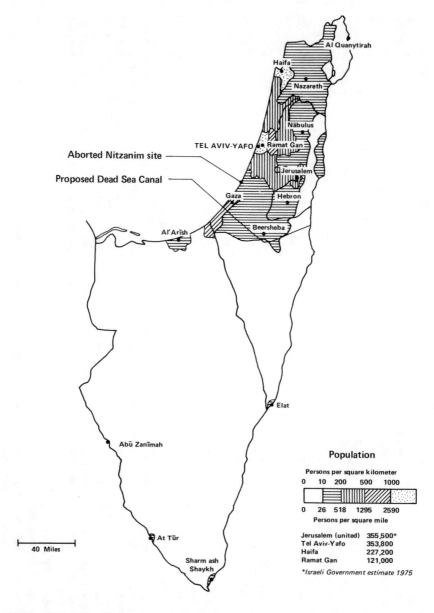

Sources: Central Intelligence Agency, ''Israel-Population,'' 54382 (Langley, Va.: Central Intelligence Agency, January 1978). *Los Angeles Times*, August 25, 1979, p. 7.

Note: It is contemplated that a nuclear power plant would be built along the canal. Neither the size nor type of proposed plant is indicated in the literature at the time of this writing.

Figure 3-3. Approximate Location of Israel's Contemplated Nuclear Power Plant

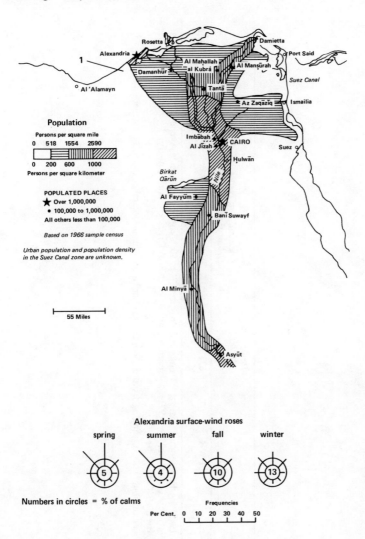

Figure 3-4. Approximate Location of Planned Sidi Kreir Reactor

Sources: "Egypt-Population," 500648 (Langley, Va.: Central Intelligence Agency, October 1971): Energy Establishment and Nuclear Power Plants Authority, Egypt, "Projected Role of Nuclear Power in Egypt and Problems Encountered in Implementation of the First Nuclear Plant," IAEA CN 36/574 (Vienna: International Atomic Energy Agency, 1977), p. 13; Office (Naval Division), Air Ministry, *Weather in the Mediterranean* (London: H.M. Stationery Office 1936), pp. 11 10-13.

Note: The seasonal surface wind flow at Sidi-Krier can be gleaned from data at Alexandria. The information is presented in the form of a wind rose. The figure in the inner circle indicates the frequency of calms, the radiating lines the direction from which the wind blows. The distance between the inner and outer circles represents a directional flow of 10 percent. Frequencies beyond the outer circle are measured as a percentile on the frequency scale.

Figure 3-4. Approximate Location of Planned Sidi Kreir Reactor

Table 3-3
Egypt Map Key

Nuclear Plant No.	Designation of Plant	Type	Power MW(e) $(100+)^a$	Location and Nearest Urban Populations[b]	Land Use in Vicinity
1	Sidi-Krier 1	PWR	622 Pl	Alexandria 2,927,000	Seasonal grazing, cotton

Sources: *Nuclear News* 22 (February 1979):69; Population Division, "Trends and Prospects," p. 50; Central Intelligence Agency, "Egypt-Land Use" 500648 (Langley, Va.: Central Intelligence Agency, October 1971); Meteorological Office (Naval Division), Air Ministry, *Weather in the Mediterranean* (London: H.M. Stationary Office, 1936) p. 11 61.

Note: The rainy season is from October through March, ranging from 61 mm in October to 66 mm in December (Alexandria reading).

[a]Pl: planned.
[b]1980 estimate.

the proposed sites is remote from population centers. If Israel constructs the Dead Sea Canal, plant winds could carry contaminants over Beersheba, a city of over 100,000. Sublethal radiation could reach Tel Aviv, assuming favorable meteorology, posing long-term health hazards. Should this occur, population relocation would be difficult because of Israel's limited geography. Furthermore a substantial portion of Israel's agricultural land would be contaminated. Even with land reclamation, consumers probably would be relucant to purchase its produce, resulting in serious economic consequences.[21]

Releases from the proposed Sidi Kreir plant might carry radioactivity toward the farming land along the Nile where population densities range from 200 to 1,000 inhabitants per km^2. Under the worst conditions, the plume would irradiate land upon which millions of people live. If relocation plans were inadequate tens of thousands would succumb to early, late, and genetic effects. At the same time the Alexandria, wind roses that appear on figure 3-4 suggest that variable winds also could blow contaminants toward uninhabited regions.

The timing of power plant construction may have significant strategic implications. Israel hopes to place nuclear power on-line in the late 1980s. Should it be the first or only country in the region to do so and should the Arabs, including the Palestinians, acquire the capability to destroy the facility, Israel's antagonists would be able to inflict widespread damage without confronting Israel's armed forces. Thus Israel would be vulnerable to a new mode of Arab intimidation. Further should radioactive products actually be released during a conflict, identification of the perpetrator might be difficult since Israel has numerous adversaries. To forestall either prospect, the Israelis might feel compelled to announce a nuclear weapons capability (which may feel it already has) for deterrence purposes.[22] The Arabs could respond with efforts to obtain their own nuclear weapons although they might be satisfied, at least in the interim, with a credible capacity to destroy atomic installations. However, this scenario might be undermined by the concurrent vulnerability of large Arab populations in Gaza and the West Bank, which could deter both threats and actions against the installation for the explicit purpose of releasing radioactive products. Still should the plant be destroyed for a variety of other reasons, the consequences would remain grave.

Should Egypt be the first to acquire a nuclear power plant, it would expose itself to manipulation not only by Israel but perhaps by some Arab states, notably Libya, whose relations with Egypt are strained. If both Egypt and Israel acquired equally vulnerable installations simultaneously, stability between the two may result, assuming that both perceive nuclear contamination to be unacceptable and believe destruction would be carried out. This would not necessarily prevent conflict for limited objectives, but it could deter either from attempting total victory. Still Israel's vulnerability and manipulation by other Arab states and parties that do not have nuclear power plants would remain. Conceivably Israel could locate nuclear wastes along the Golan Heights and perhaps along its border with Jordan or some future Palestinian entity. Assuming reliable westerly winds, the nuclear materials could function for either deterrence or defense. However, the possibility of Israeli self-contamination makes this option unrealistic. Furthermore other Arab antagonists that do not border Israel, notably Iraq and Saudi Arabia, would have to be sensitive to the fate of their Arab allies to be deterred. Similar scenarios would apply in cases where other nations in the region acquired nuclear energy.

The implications of facility vulnerability and its impact on stability go beyond the Middle East itself because of the economic and political interests of the United States and the Soviet Union. The United States could find that Israeli vulnerability forces it to play a more active role to deter threats and to reassure Tel Aviv that asymetrical possession does not require a compensatory nuclear weapons declaration. In the longer run should Soviet client states such as Syria acquire nuclear facilities, they might want a greater commitment of Soviet support against Israeli intimidation. Greater involvement of either superpower raises the risk of future confrontation.

Asia and Africa

The scenarios suggested for Europe and the Middle East can apply to at least four other regions: the Korean peninsula, South Asia, West Asia, and southern Africa.

Korea

Korean partition into Soviet and American sectors after World War II has been a continual source of international tension. Although no major conflict has erupted since the end of the Korean War in 1953, a number of small skirmishes keep alive speculation that war is an ever-possible result of either miscalculation or an attempt by either side to use force to reunite North and South Korea.[23] In the event war is threatened or erupts, the presence of nuclear power plants may play a significant role.

It is strategically significant that only South Korea has a nuclear energy program with one reactor into operation, four others under construction, and two on order.[24] As figure 3-5 indicates, all three are located in the southern tip of the country over two hundred miles from the North Korean border in a region with moderate population density (100 to 200 persons km²). (See table 3-4.) In the long term the South Koreans plan an additional forty plants.[25]

The distance of reactors from the border helps diminish their vulnerability to military destruction, but should the North be able to penetrate the South's defenses—facilitated should future South Korean facilities be situated closer to the border—significant strategic implications could arise. The North could threaten the facilities for intimidation purposes, or if the plants were located near military concentrations, they could be destroyed as part of a military strategy. Although it could improve its defenses, Seoul could not respond in kind because the North does not possess nuclear energy. Furthermore it is questionable whether the South could readily use facilities as defensive or deterrent radiological frontier mines because of seasonal changes in prevailing monsoonal winds, which flow from the north in winter and south in summer. However, as the installations are currently located, weather could help diminish the threat to the South. Heavy summer rainfall would localize releases. From October to April when rains are the lightest, prevailing northerly winds would carry products out to sea, although releases probably would be subject to onshore sea breezes.

South Korean concern about the manipulative implications of its installations could contribute to ever-increasing insecurity if the United States withdraws its ground forces in the 1980s and the American defense commitment becomes less credible. The South's disposition to acquire nuclear weapons may correspondingly increase. To forestall this possibility, the United States might place greater reliance on nuclear weapons to deter North Korea. Or the South might sense that the United States would be more cautious if the North attained a credible ability to release the radionuclides. In any case, asymmetries in the peninsula's nuclear energy development could be destabilizing.

China and Taiwan

The Chinese context has similarities to the Korean but also has significant differences. Like Korea, a divided nation confronts itself where nuclear power development is asymmetrical. At the present time, only Taiwan operates nuclear energy plants; it has two 600 MW(e) reactors on line and four 900 MW(e) plants under construction. (See figure 3-6 and table 3-5.) By contrast, the plans of the People's Republic are uncertain. In 1978 it announced it would purchase two French reactors; in 1979 it canceled the agreement. Early in 1980 it announced plans for a 900 MW(e) plant in Shenzhen near Hong Kong.[26]

It is uncertain whether Taiwan's growing nuclear program will

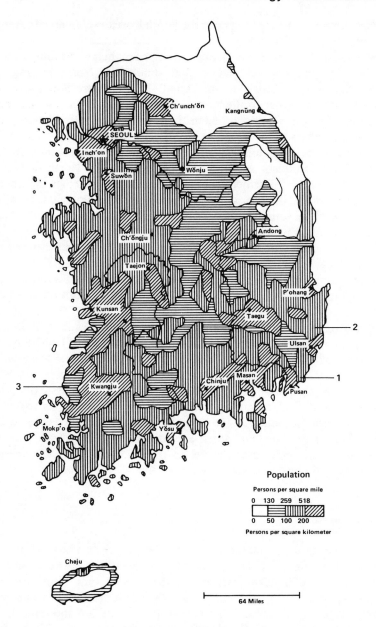

Sources: Central Intelligence Agency, "South Korea-Population," 501379 (Langley, Va.: Central Intelligence Agency, September 1973): "World List of Nuclear Power Plants," *Nuclear News* 22 (August 1979): 74; Times, *The Times Atlas of the World: Comprehensive Edition* (London: Times Newspapers Ltd., 1967) Plate 21.

Figure 3-5. Approximate Locations of South Korea's Power Reactors

Table 3-4
South Korea Key Map

Nuclear Plant No.	Designation of Plant	Type	Power MW(e) (100+)[a]	Location and Nearest Urban Populations[b]	Land Use in Vicinity
1	Ko-Ri 1	PWR	564	Pusan City 2,857,000	Forest
	Ko-Ri 2	PWR	605 U/C		
	Ko-Ri 3	PWR	994 U/C		
	Ko-Ri 4	PWR	994 U/C		
2	Wolsung 1	PHWR	620 U/C	Kyong Ju Ulsan City 238,000	Forest, scrub, brush, barren land, paddy crops
3	Korean nuclear unit 7	PWR	900 Or	Mokp'o 113,000	Paddy crops, dry crops, forest
	Korean nuclear unit 8	PWR	900 Or		

Sources: "World List of Nuclear Power Plants," p. 74; "Westinghouse Fights for Philippines Order But Wins in Korea," *Nuclear Engineering International* 24 (September 1979):3; Atomic Industrial Forum, *INFO* (Washington, D.C.: Atomic Industrial Forum, February 6, 1980), p. 7; Population Division, "Trends and Prospects," p. 45; Editors of Life and Rand McNally, p. 532; Central Intelligence Agency, "Korea-Land Use," 501379 (Langley, Va.: Central Intelligence Agency, September 1973); Bartholomew et al., *Atlas*, plates 12, 20.

Notes: Prevailing winds for the winter monsoon are northwest and for the summer monsoon, southerly. Precipitation for April through September averages two to eight inches per month and is heaviest in the south. During the remainder of the year, it is less than two inches per month.

[a]UC: under construction; Or: ordered.

[b]All population data are 1980 estimates, except for Mokp'o, which is a 1960 estimate.

significantly change the strategic balance between itself and Peking. Unlike South Korea, it already faces a nuclear weapons threat from its antagonist. Certainly the implications of facility vulnerability should be pondered. However, it is difficult to judge to what extent this vulnerability gives the People's Republic greater leverage. Indeed Taipei could use the issue to legitimize its acquisition of nuclear weapons as a countermeasure.

Should the People's Republic acquire nuclear energy, it too could face problems from several quarters. Certainly Taiwan and perhaps India, with which Peking's relations are cool, might find the facilities attractive targets for purposes of deterrence or coercive diplomacy. However, their ability to manipulate the plants would be limited by great distances and inadequate military ordnance. More credible threats could come from South Korea and Japan and perhaps Vietnam. The nuclear energy deterrent in these cases could serve to stabilize the military balance between them and China. Finally both the Soviet Union and the United States could destroy Chinese facilities. However, it is unlikely they would gain any additional strategic leverage given their already vast nuclear weapon superiority over Peking. Nonetheless, such destruction would make China's postwar recovery more difficult. It is questionable whether the Chinese could credibly counter the Soviet threat with facilities on their common frontier because of variable winds.[27]

India and Pakistan

In South Asia, a somewhat different asymmetry in nuclear development exists between India and Pakistan. Figure 3-7 shows that India operates four 200 MW(e) reactors near Bombay and Kota and is building four slightly larger ones in other agricultural regions, with population densities ranging from 100 to 200 km^2 (see table 3-6). In addition, India has built small reprocessing and fuel fabrication facilities. Another important development in India's atomic program was the 1974 detonation of an underground nuclear device. By contrast, Pakistan operates one small 125 MW(e) reactor near Karachi and plans one other but is also in the market for a reprocessing plant and is planning an enrichment facility.

Since the partition of Pakistan and India upon independence from Britain in 1947, relations between the two have been strained due to territorial disputes, strong religious animosities, and national ambitions.[28] Wars were fought in 1947-1949 and 1965 principally over Kashmir, with the last conflict including fighting along the Punjab border and the Ran of Cutch. The most significant clashes occurred in 1971 when in the midst of West Pakistan's repression of East Pakistan's efforts at autonomy, war broke out, resulting in India's occupation of Dacca and the creation of

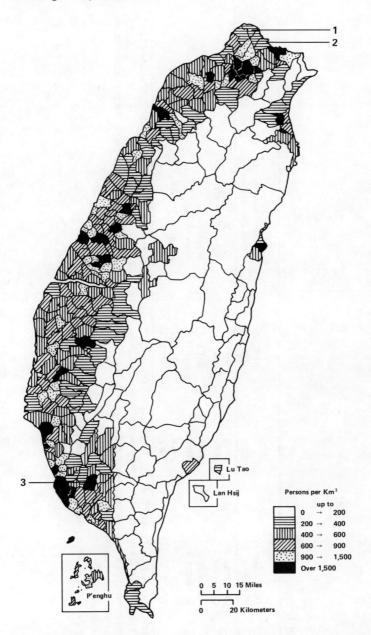

Sources: Reproduced by the *Annals* of the Association of American Geographers Map Supplement Series, No. 11, "Density of Population," 1965, Vol. 59. No. 3, 1969, Mei-ling, Hsu: Times.

Figure 3-6. Approximate Location of Taiwan's Power Reactors

Table 3-5
Taiwan Map Key

Nuclear Plant No.	Designation of Plant	Type	Power MW(e) (100+)[a]	Location and Nearest Urban Populations[b]	Land Use in Vicinity
1	Chin-shan 1 Chin-shan 2	BWR BWR	604 604	Shihmin Hsian Taipei 2,377,000	Rice
2	Kuosheng 1 Kuosheng 2	BWR BWR	951 U/C 951 U/C	Wanli Hsian Keelung 395,000	Rice
3	TPC 1 TPC 2	PWR PWR	907 U/C 907 U/C	Pingtung 110,000	Rice

Sources: "World List of Nuclear Power Plants," p. 76; Population Division, "Trends and Prospects," p. 56; Meil-Ling Hsu, "Taiwan Land Use" (Minneapolis, Minn.: University of Minnesota, 1969); Bartholomew et al., *Atlas*, plates 12, 20.

Notes: Prevailing geostrophic wind is northeast except in the summer when it is westerly. Precipitation in winter is two to four inches per month; the remainder of the year it is four to twelve inches per month.

[a]U/C: under construction.

[b]Population data are 1980 estimates, except for Pingtung, which is a 1960 estimate.

Bangladesh. Creation of the new state led to West Pakistani concern about New Delhi's use of subcontinental predominance. India's 1974 nuclear detonation fueled those concerns, and Pakistan began to consider the nuclear weapons option.[29] Pakistan's pursuit of the option may lead to an arms race, which in turn would increase suspicions, tensions, and the possibility of war.

Targeting nuclear energy facilities might offer a means to avoid this dilemma. Consistent with India's ability to develop nuclear weapons, Pakistan might threaten to acquire reliable advanced weapons systems that could destroy the installations. Indeed Pakistan's fleet of eleven light Canberra bombers may constitute a credible attack force now.[30]

Both India and Pakistan face outside dangers. Pakistan has border disputes with Afghanistan and trouble with ethnic groups that are seeking autonomy. However, as long as Pakistan's nuclear energy program remains modest, it is unlikely to have any strategic wartime implications. India must concern itself with the People's Republic, with which it fought a border war in 1962. Given China's conventional weapons superiority and nuclear weapons capability, as well as the distance of Indian facilities from its border, Peking probably would not consider the installations priority targets. Variable winds along this frontier, as well as along the Indo-Pakistani border, make the use of nuclear installations as border mines impractical.[31]

West Asia

The future of nuclear energy in West Asia, a region roughly stretching from Afghanistan to the Persian Gulf, is uncertain. Only Iran has begun construction of two 1,200 MW(e) PWRs near Büshehr, a Persian Gulf city with approximately 27,000 inhabitants (1960 estimate) surrounded by a sparsely populated region with sporatic agriculture and grazing. Under the shah, a number of additional reactors were planned. But with his overthrow in 1979, plans have been scuttled, and construction of the Büshehr installations has stalled.[32] It remains to be seen if and when the Persian Gulf plants will be completed. (See figure 3-8 and table 3-7.)

Should Iran complete the Büshehr construction and build additional plants, their existence may have strategic implications in times of tensions and conflict between Tehran and some of its neighbors. Surrounded by Pakistan, Afghanistan, the Soviet Union, Turkey, Iraq, and Saudi Arabia as well as several sheikdoms and smaller countries along the Persian Gulf, Iran confronts a complex security problem, but one it managed rather well, at least under the shah.[33] Until the early 1960s, Soviet-Iranian relations were strained, but Russian attempts to undermine the country were resisted successfully. In the late 1940s, Iran, with the help of the United States, induced

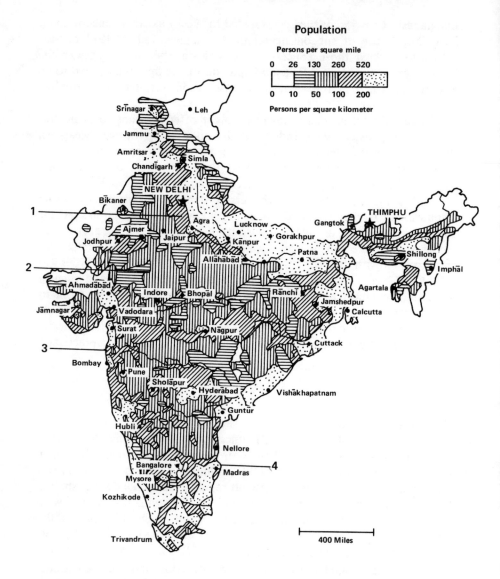

Population

Persons per square mile

0   26   130   260   520

0   10   50   100   200

Persons per square kilometer

Srinagar   • Leh

Jammu

Amritsar

Chandigarh   Simla

NEW DELHI

Bikaner

1

Ajmer   Agra   Lucknow   Gangtok   THIMPHU

Jodhpur   Jaipur   Kanpur   Gorakhpur   Shillong

Allahabad   Patna   Imphal

2

Ahmadabad   Agartala

Jamnagar   Indore   Bhopal   Ranchi   Jamshedpur

Vadodara   Calcutta

Surat   Nagpur

3   Cuttack

Bombay   Pune

Sholapur   Vishakhapatnam

Hyderabad

Guntur

Hubli

Nellore

Bangalore   Madras   4

Mysore

Kozhikode

Trivandrum   400 Miles

Sources: Central Intelligence Agency, "India-Population," 501957 (Langley, Va.: Central Intelligence Agency, July 1973); "World List," pp. 72-73; Times, *Times Atlas*, plates 28, 29.

Figure 3-7. Approximate Location of India's Power Reactors

Russian withdrawal from territory occupied during World War II and overcame Soviet-sponsored secessionist movements. In the 1950s, it withstood

Table 3-6
India Map Key

Nuclear Plant No.	Designation of Plant	Type	Power MW(e) (100+)[a]	Location and Nearest Urban Populations[b]	Land Use in Vicinity
1	NAPP 1 NAPP 2	PHWR PHWR	220 U/C 220 U/C	Narora New Delhi 5,704,000	Rice, millet
2	RAPP 1 RAPP 2	PHWR PHWR	202 202	Kota 65,000	Rice
3	Tarapur 1 Tarapur 2	BWR BWR	200 200	Bombay 8,722,000	Rice
4	MAPP 1 MAPP 2	PHWR PHWR	220 U/C 220 U/C	Kalpakkam Madras 5,704,000	Rice

Sources: "World List of Nuclear Power Plants," pp. 73-74; Atomic Industrial Forum, *INFO*, pp. 5-6; Population Division, "Trends and Prospects," pp. 47-49; Central Intelligence Agency, "India-Land Use," 501957 (Langley, Va.: Central Intelligence Agency, July 1973); Bartholomew et al., *Atlas*, plates 12, 25.

Notes: Prevailing winds for the winter monsoon are northeast and for the summer monsoon, southwest. Precipitation is greatest June through September, ranging from four to over sixteen inches per month over most of the country and where nuclear facilities are situated. From December through May, most of the country receives under one inch per month.

[a]U/C: under construction.

[b]All population data are 1980 estimates, except for Kota, which is a 1960 estimate.

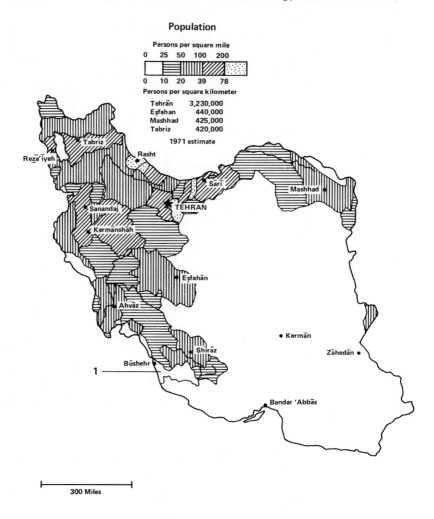

Population

Persons per square mile

| 0 | 25 | 50 | 100 | 200 |

| 0 | 10 | 20 | 39 | 78 |

Persons per square kilometer

Tehrān	3,230,000
Eşfahan	440,000
Mashhad	425,000
Tabriz	420,000

1971 estimate

300 Miles

Sources: Central Intelligence Agency, "Iran-Population," 501163 (Langley, Va.: Central Intelligence Agency, March 1973); "World List," p. 73; Times, *Times Atlas*, plate 32.

Figure 3-8. Approximate Location of Iran's Power Reactors

Soviet intimidation, including border incidents, airspace violations, and espionage. By the 1960s, however, both states sought accommodation—Iran to limit its excessive dependence on the United States and the Soviet Union to improve relations with the rest of Asia as it became preoccupied with China. Trade expanded. The Russians provided technical aid but remained neutral in disputes Tehran had with other neighbors. Today relations are correct

Table 3-7
Iran Map Key

Nuclear Plant No.	Designation of Plant	Type	Power MW(e) (100+)[a]	Location and Nearest Urban Populations[b]	Land Use in Vicinity
1	Iran 1 Iran 2	PWR PWR	1,200 U/C 1,200 U/C	Bushehr 27,317	Sporadic agriculture and grazing

Sources: "World List of Nuclear Power Plants," p. 73; Editors of Life and Rand McNally, p. 460; Central Intelligence Agency, "Iran-Land Use," 501163 (Langley, Va.: Central Intelligence Agency, July 1973); Bartholomew et al., *Atlas*, plates 12, 20.

Notes: Prevailing geostrophic winds are north-northwesterly most of the year. The region where the reactor is located receives two inches or less per month from November through March. It is dry the remainder of the year.

[a]U/C: under construction.

[b]1980 estimate.

but not warm. Iran is concerned about the portent of Soviet activity in Afghanistan, the Indian Ocean, and Moscow's support of radicals attempting to overthrow established regimes in the gulf and separatist movements in Pakistan whose success could motivate kindred communities in Iran.

Along its western frontier, Iran's principal challenges come from Iraq. Prior to 1958 the two countries were aligned in the Baghdad Pact, but in that year the Iraqi monarchy was overthrown and succeeded by a radical regime. In 1959, the Iraqis initiated a crisis over the Shatt-al-Arab, a marine shipping channel whose sovereignty had never been established. Consequently other disputes arose, including one over Iraq's support of revolutionary movements in the Persian Gulf and Iran's support of Kurdish rebels in Iraq. Deteriorating relations during the 1960s and early 1970s led to serious border clashes. In 1975 negotiations culminated in an agreement ending the waterway dispute and Iran's support of the Kurds. This resulted in a lessening of tensions and ended a fifteen-year Iranian effort to ensure the Persian Gulf developed in a direction consistent with its strategic interests. However, with the overthrow of the shah relations have once again become unsettled with border clashes reported in the press.

Simultaneously, during the mid-1960s, Egypt seemed intent on overthrowing the region's conservative regimes. Nasser committed large forces to support the antiroyalist cause in Yemen's civil war, an act Iran viewed as a precursor to an Egyptian military threat. However, relations improved significantly after Cairo withdrew its troops in 1967 and Nasser died in 1970. Although they have deteriorated with the overthrow of the shah and Iran's opposition to Egypt's peace efforts with Israel, this is unlikely to lead to a military confrontation.

The potential for conflict between Iran and its smaller neighbors did not cease with the end of the Nasser period. In 1968 Great Britain declared its intention to withdraw its military forces from the Persian Gulf by 1971. Iran retained claims to British-controlled Bahrain and Abu Musa and the Greater and Lesser Tumbs, small islands near the Strait of Hormuz, the vital passage to the Indian Ocean. To ensure that the islands did not fall into unfriendly hands, Iran occupied them. A number of Arab sheikdoms also were concerned about how Tehran would treat them once the British departed. To allay fears of Iranian imperialism, Iran relinquished its claim to Bahrain and withdrew objections to the formation of a union of Arab emirates. At the same time, Iran kept a close watch to avert radicalism in the region, demonstrating its commitment by sending several thousand troops to Oman to put down the Dhofari insurgency.

Finally, along Iran's eastern frontier the principal threat comes from Afghanistan's support of Iranian seccessionist movements and the portent of the Soviet presence.

The implications of Iranian nuclear energy facilities for stability in West Asia will depend on future political developments. Should Tehran seek a more dominant role over the affairs of the smaller gulf states or acquire nuclear weapons either for purposes of security or prestige, its nuclear facilities might afford these nations a radiological deterrent, assuming they had the weapons to destroy the installations. At the same time, Iran itself could be vulnerable to intimidation, although threats might be diminished by the fact that winds would carry products into uninhabited regions. It seems unlikely that the power plants would play a role in relations with the Soviet Union because of the Soviet Union's military superiority.

South Africa

What level of violence might be used by South Africa's indigenous blacks, and expatriots and their supporters abroad to end apartheid? To date it has been limited to several large riots in black townships and acts of terrorism in the country at large. Many observers forecast increased guerrilla activity.[34]

To meet this challenge as well as any conventional military threat, South Africa has assembled the best military force in Africa, consisting of 65,000 men, which can expand to 404,500 when totally mobilized. It is equipped with over 250 tanks, 3 submarines and 1 destroyer, 345 combat aircraft including modern interceptors and bombers, and an early warning system to detect aerial intrusions.[35] The South Africans can build nuclear weapons and may have tested them already.[36] South Africa has a well-financed and manned nuclear research program centered at Pelindaba near Pretoria where two research reactors are located, the largest being 20 MW(e). (See figure 3-9 and table 3-8.) A small enrichment plant operates in nearby Valineaba, and there are plans to build a large commercial facility. Also under construction are two 922 MW(e) PWRs 30 km north of Capetown in Koeberg whose principal purpose, some have speculated, is the production of plutonium for a weapons program.[37]

South Africa's facilities are likely to be attractive military targets because they appear to be a guise for a nuclear weapons program, they have considerable intrinsic value, and in the case of the future Koeberg facilities, they will generate electrical power with potential to contaminate large regions if damaged or destroyed. At the same time, the installations are difficult to destroy because of their internal locations and because of the limited military capabilities of African adversaries. The nearest potential adversary is Mozambique, located approximately 250 mi. from Pelindaba and 900 mi. from Koeberg. Certainly no African country today could attack these facilities with conventional armed forces given South Africa's defenses. Commandos might have a better chance, but the task would still

Population

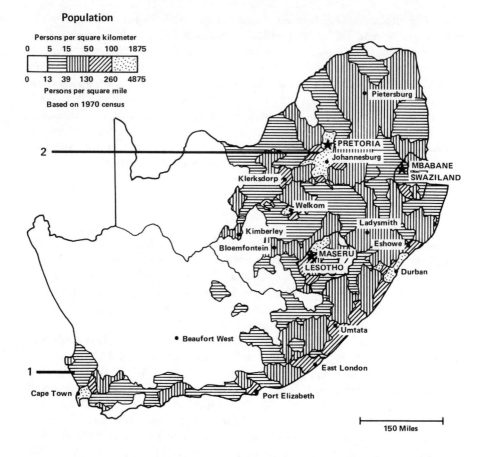

Source: Central Intelligence Agency, "South Africa-Population," 503971 (Langley, Va.: Central Intelligence Agency, April 1979).

Figure 3-9. Approximate Location of South Africa's Nuclear Installations

be difficult because the installations are well guarded. Only foreign troops, such as Soviet bloc forces equipped with sophisticated aircraft or missiles, could launch a damaging attack.

If the installations could be successfully attacked, they might afford South Africa's adversaries a nuclear hostage assuming the white population was sensitive to the consequences. The possibility that releases would irradiate the black population or, in the case of the Koeberg facility, be carried by prevailing southerly winds into relatively uninhabited regions could diminish the impact. If the white population were concerned, threats to destroy the facilities could deter major South African incursions into

Table 3-8
South Africa Map Key

Nuclear Plant No.	Designation of Plant	Type	Power[a]	Location and Nearest Urban Populations[b]	Land Use in Vicinity
1	Koeberg 1 Koeberg 2	PWR PWR	922 U/C 922 U/C MW(e)	Cape Town 1,545,000	Mediterranean agriculture
2	Safari 1	Research	20 U/C MW(th)	Pelindaba Pretoria and Johannesburg 3,933,000 Johannesburg 2,772,000	General farming
	Pelunduna Zero UCOR	Research negligible Enrichment	Pilot and commercial U/C	Valindaba Pretoria and Johannesburg	

Sources: "World List of Nuclear Power Plants," p. 75; Population Division, "Trends and Prospects," p. 35; Editors of Life and Rand McNally, *Atlas*, p. 422; Bartholomew et al., *Atlas*, plate 12.

Notes: Prevailing geostrophic winds for Koeberg are southeast in January and southwest in July. For Pelindaba and Valindaba, they are northwest in January and July. The rainy season at Koeberg begins in April and ends in September, averaging two to six inches per month. During the remainder of the year, precipitation is under two inches per month. At Pelindaba and Valindaba the rainy season begins in October and extends through April, ranging from two to over six inches per month. During the remainder of the year, precipitation is under two inches per month.

[a]U/C: under construction.

[b]1980 estimates.

neighboring states such as the invasion of Angola in recent years. It could deflate the impact of South Africa's nuclear weapons capability. In time of crisis, the black African states and their supporters could threaten the facilities for coercive diplomatic purposes. And if the facilities were near military installations, they could be threatened or destroyed as part of military strategy. South Africa's own ability to use installations as radiological mines would be limited due to variable winds along the borders.[38] In summary, such circumstances could enhance stability. However, vulnerability could incite South African belligerence; South Africa could pursue a nuclear weapons program openly to demonstrate its resolve not to be manipulated. These scenarios remain speculative as long as South Africa's adversaries are poorly armed, which is likely for the foreseeable future, and powerful foreign forces stay out of the region.

United States

The United States has 191 reactors built, under construction, or planned, in addition to several fuel fabrication facilities, reprocessing facilities, and high-level waste storage areas.[39] When the Chesters examined the wartime vulnerability of the American nuclear energy industry through the end of this century, they concluded that the Soviets are likely to acquire increasingly lethal nuclear weapon capability over U.S. facilities, a large fraction of which, if ground-burst, could add significantly to the radioactive fallout, perhaps doubling the cancer rate due to residual radiation exposure in the postattack environment. However, these effects are relatively minor:

> A few hundred thousand additional cancer cases per year, starting a decade or more after the attack, from targeting the nuclear industry would be difficult to detect against the background of a total of 100 to 120 million immediate fatalities and from weapon effects and the cancers induced by the large fallout radiation absorbed by the survivors.[40]

Even after 100 million deaths, hundreds of thousands of additional cancers per year, even ten years after a conflict, is not something to take lightly. Nonetheless given the magnitude of the nuclear weapons threat, the Soviet ability to release facility products is not likely to give them any additional leverage over the United States for purposes of coercive diplomacy, although it might have deterrence value if the Russians are really concerned about a belligerent America. Militarily it would complicate any operations following bombardment and, above all else, complicate postwar recovery. This last fact is reflected in the contamination contours in figures 3-10 and 3-11. Figure 3-10 indicates the gamma dose rate one year after a nuclear weapons attack excluding the nuclear industry in the year 2000. Figure 3-11

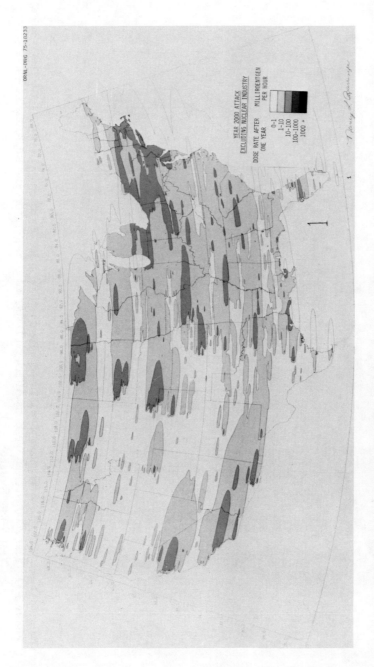

Source: Conrad V. Chester and Rowena O. Chester, "Civil Defense Implications of the U.S. Nuclear Power Industry During a Large Nuclear War in the Year 2000," *Nuclear Technology* 31 (December 1976):335.

Figure 3-10. Year 2000 Nuclear Weapons Attack, Excluding U.S. Nuclear Industry

Source: Chester and Chester, "Civil Defense Implications of the U.S. Nuclear Power Industry," p. 334. (For elaboration See Figs 9 and 10 in Chesters' article.)

Figure 3-11. Year 2000 Nuclear Weapons Attack on Dispersed U.S. Reactors and Reprocessing Plants—Ten-Year Waste Storage

assumes the destruction of existing, planned, and hypothetical nuclear facilities, including reprocessing facilities containing ten-year waste storage. The greater long-term contamination of the second scenario is readily apparent.

Conclusions

Successful manipulation of nuclear facilities for deterrence, coercive diplomacy, and military strategy requires fulfillment of a number of criteria. First, combatants must have credible ability to release radionuclides into the environment. This credibility is not a problem for nations using their own facilities as defensive or deterrent radiological mines; it will be for belligerents attacking the facilities of others. The successful attacker must penetrate active defenses and disrupt an installation's safety features. At this time the United States, the Soviet Union, and probably Great Britain, France, East and West Germany, and Israel could destroy the installations of their respective antagonists. It is less certain whether North Korea, Pakistan, the Arab states, and China could do so given their current capabilities and their adversaries' defenses. It is unlikely that Yugoslavia or Iran (assuming disrepair of its armed forces) could penetrate Soviet defenses or that black Africa could destroy South Africa's facilities. However, in some of these less certain and doubtful cases where destruction through bombardment would be difficult, terrorist surrogates or commandos might be more effective.

Whether destruction results in severe consequences depends upon the location of installations in relation to population, valued land, and military concentrations, as well as upon weather and the functioning of emergency systems designed to mitigate releases. From the deterrence and coercive diplomacy perspective, releases optimally must contaminate heavily populated and intensely farmed land. In some cases this criterion is better met than are others. Indian reactors are located near the country's densely populated western coastal region. Cities are built around Soviet facilities. Any release of radionuclides in Israel would be catastrophic given the country's small size; the same probably could be said for the West European countries. By contrast, Iranian plants under construction in Büshehr are situated in a remote region; thus any consequences would not be as great. Militarily facility location is relatively more important, and successful attack requires some precision. Thus to be effective, troop concentrations must be downwind from a facility or combatants must cross territory that can be contaminated, such as the invasion routes into West Germany.

Weather always plays an important role in the timing of war and in its conduct. It could also be significant in determining when facilities should be attacked. Obviously the presumption of stable weather, drawn from Beyea and used as a reference point in some of the case studies, does not reflect numerous variables to which radionuclides could be subject. Consequence models show that turbulence will dilute contaminants; precipitation at the time of a release will localize the impact around the facility; products lifted into rain-bearing clouds can carry early lethal products one hundred miles; weather inversions (common in the evening) will extend the lethal plume; and winds, often affected by local geography, will carry materials away from rather than toward populations and valued land.

Although weather is fickle, general predictions can be made and used for planning attacks. For example, in central Europe and European Russia, precipitation is greatest from May through August; therefore releases are more likely to be washed out during this period than during other times of the year. In Korea, the situation is more complicated. Winds from the south that could carry contaminants over South Korea from facilities located at the tip of the peninsula occur in the summer. However, they also bring with them South Korea's heaviest rains. Winter winds come from the north; these could carry materials out to sea. In Israel, contaminants subject to prevailing southwesterly coastal winds could spread to Israel's most populated regions; attacks would probably be most effective from April or May through October when it does not rain.[41] Egypt has a shorter rainy season and northwesterly winds flow much of the year.

The effectiveness of facility safety features, such as sprays and ice condensers in reactor containment vessels, will affect contamination. The more redundant the safety features, the greater the probability of on-site containment of releases. This point underscores the greater danger posed by the Soviet reactors, which have fewer safety features than those built in the West.

The ultimate test of facility manipulatability is the target state's psychological sensitivity to the threat of nuclear contamination. Sensitivity is a product of different cognitive factors including rationality plus estimates of the physical consequences and faith in countermeasures. Leaders will approach this matter differently Some may not consider the problem significant. They will argue that numerous variables that act on facility radiation reduce its effectiveness. Such optimism could be justified if population centers and valued land are not in the vicinity of the facility, installations are downwind from valued locations, meteorology unusually promotes the rapid wind dilution or washing out of materials, facilities have numerous safety features, regard for adversary military capabilities is low, adequate shelters are available and relocation plans devised, and/or contamination estimates are predicted to have few immediate effects and long-term consequences can be dealt with reasonably. Also wishful thinking supports optimism along all of these lines, reinforced by the relative unreliability of facility radiation compared to that of nuclear weapons. On the other hand, some leaders are hypersensitive to the prospect of any nuclear contamination. Those countries most sensitive to the dangers of accidental releases could be the most manipulatable and the most disposed to take drastic measures such as acquisition of nuclear weapons to change the strategic balance. The reaction spectrum is wide. In addition, potential aggressors might overestimate the manipulatability of facilities for coercive diplomacy or even military strategy, thus taking actions they might not have contemplated otherwise.

Given the absence of debate in all countries about the implications of facility vulnerability, specific response for each entity is impossible to determine. However, claims, such as those by Chester Cooper, that civil nuclear contamination could function as nuclear weapons have in stabilizing the strategic balance between the United States and the Soviet Union should be scrutinized very carefully (as indeed Cooper has cautioned), taking into account the numerous variables that affect such calculations. The superpower example does not necessarily prove the case for stability, although it has been used by advocates of nuclear weapons proliferation to the Middle East.[42] However, the superpower relationship is based on a largely unwritten code of conduct that has evolved by mutual and tacit consent since World War II. This code has been reinforced by lengthy consultations and by symmetries in military development. Even with this code, the world was brought to the brink of nuclear war during the 1962 Cuban missile crisis. While it may be presumptuous to believe that other antagonists could not act with the restraint shown by the Russians and Americans, neither should it be assumed that they will.[43]

4

Options to Diminish the Wartime Vulnerability of Nuclear Energy Facilities

The politically destabilizing implications of nuclear facility wartime vulnerability should stimulate the search for alternatives that enhance stability. But an equally compelling stimulant is the immorality of holding millions of civilians and unborn generations hostage to the ravages of cancer, genetic disease, and associated psychological traumatization. Fred Iklé's critique of American nuclear weapon targeting doctrine is relevant. Iklé argues that sensitivity to the distinction between combatants and civilians cultivated through the centuries was dulled by the strategic bombing campaign of World War II. In the nuclear era, military planners have become increasingly insensitive to the implications of holding millions of Russians hostage "by layers of dehumanizing abstractions and bland metaphors."[1] Such jargon as *assured destruction, unacceptable damage,* and *deterrence* generally works on strategic analysis like "a narcotic":

> It dulls our sense of moral outrage about the tragic confrontation of nuclear arsenals, primed and constantly perfected to unleash widespread genocide. It fosters the current smug complacence regarding the soundness and stability of mutual deterrence. It blinds us to the fact that our methods of preventing nuclear war rests on a form of warfare universally condemned since the Dark Ages—the mass killing of hostages.

> Indeed, our nuclear strategy is supposed to work better, the larger the number of hostages that would pay with their lives should the strategy fail. This view has become so ingrained that the number of hostages who could be killed through a "second" strike by either superpower is often used as a measure of the "stability" of deterrence.[2]

Iklé calls for greater discrimination in American targeting doctrine made possible by increasingly accurate delivery systems. To resolve this predicament further in the context of nuclear energy while at the same time encouraging stability, I present three nonexclusive sets of alternatives. The first involves controlling national behavior with international law. The second proposes physical means to reduce or prevent consequences of facility destruction; these include civil defense, different modes of siting facilities and improving their inherent safety, and adoption of alternative energy sources. The third suggests how international institutions can manage nuclear exports and indigenous production to minimize wartime dangers.

113

A Proposal for Legal Restraint

I begin with legal restraint not because it is the most reliable option to minimize the military threat posed to nuclear energy (it is not) but because it offers a relatively expeditious as well as inexpensive means to address the problem by establishing a standard of behavior where one does not now exist. This may offer some modicum of restraint, which in time can be supplemented by other more authoritative alternatives that require longer lead times for implementation.

International law prescribes norms of international conduct applied to conflict that "attempts to reconcile minimum morality with the practical realities of war."[3] This reconciliation covers the commencement of war, the conduct of hostilities, neutral rights, the conclusion of combat, and the reconstruction and rehabilitation of combatants and their territories.[4] Sources of international law include conventions and treaties and what the statute of the International Court of Justice refers to as "the general principles of law recognized by civilized nations."[5] Subsidiary means of determining the rules of law include judicial decisions, works of acknowledged scholars, and perhaps resolutions by the U.N. General Assembly and interpretations by the International Committee of the Red Cross.[6]

International law inadequately addresses the destruction of nuclear energy facilities in war. Existent treaties and practices leave doubt as to whether such an action is acceptable conduct. An examination of deficiencies follows, exploring the utility of codification, and suggesting a treaty to outlaw the destruction of nuclear energy installations in war.

Existing International Conventions

Applicability of Precedents Relating to Poisons. International law does not specifically treat the legitimacy of using radiological agents in war, but what it says about the use of poisons may be relevant. However, some questions that arise concern whether radiation can be properly defined as a poison and whether a specific prohibition can be applied to an analogous case.

There are internatonal agreements concerning poisons, but none defines the term clearly. The two Hague Conventions Respecting the Laws and Customs on War on Land (1899 and 1907) are twentieth-century foundations of the laws of war.[7] Both explicitly prohibit the use of poison or poisoned weapons.[8] The 1899 convention also prohibits "the use of projectiles the object of which is diffusion of asphyxiating or deleterious gases."[9] The Protocol for the Prohibition of Poisonous Gases and Bacteriological Methods of Warfare (1925) codifies a similar prohibition.[10] In 1971, the Conference of the Committee on Disarmament negotiated an accord that prohibited the development, production, or stockpiling of bacteriological and toxin weapons and mandated the destruction of existing accumulations.[11] Subsequent General Assembly resolutions have reaffirmed international support for these conventions. However, none of them gives a clear, all-inclusive definition of *poison*.[12]

Ambiguity in treaty law and international opinion allows room for varying interpretations. Schwarzenberger, in a study devoted to the legality of nuclear weapons, contends that radioactive substances released in nuclear explosions may be defined as poisons:

> Consultation of standard textbooks and military manuals on the meaning of poison and poisoned weapons further confirms how little explored this field is. Definitions excel by their absence. Thus, no choice exists but to fall back on the ordinary meaning of these words. Etymologically, the determination of "poison" from *potio* might suggest a limitation to substances which are transmitted in fluid form. Yet, a mere glance at the Latin and German equivalents of the term makes it difficult to pay much attention to this argument. In any case, in accordance with the general rules of international law, the present day meaning of these terms is ultimately decisive. In contemporary usage, the term covers—not only in English—any substance that "when introduced into, or absorbed by, a living organism destroys life or injures health." This excludes death or injury to health by means of force, whether the cut of a sword, the thrust of a spear, the piercing of the body by an arrow or bullet, or injury inflicted by explosion or blast. Thus, it is little in doubt that the blast effects of nuclear weapons are not covered by the prohibition of the use of poison.
>
> It would be less justified to dismiss similarly out of hand the possibility that the rule on the prohibition of the use of poison might govern at least the heat and radiation effect of their fall-out. It is probably permissible to treat heat and radiation as substances. If any of these substances were introduced into the body in sufficiently large doses, they would destroy life or injure health. In any case, this is true of radioactive fall-out. Thus without invoking even the possibility of the necessary production of poison gas by nuclear explosions, a *prima facie* case apears to exist for regarding the use of nuclear weapons as incompatible with the prohibition of the use of poison.[13]

Can the same reasoning be applied to radionuclides released by the destruction of nuclear energy facilities? Other scholars, such as McDougal and Feliciano, contest Schwarzenberger's argument by analogy on the grounds that treaties should be assumed to cover only the issues that occupied the negotiators' attention:

> The assumption which may be seen to underlie . . . exercises in analogical interpretation by Dr. Schwarzenberger and others are that words have absolutistic meaning which can be projected into the future without regard to original and contemporaneous contexts, and that future interpreters must

accept these pristine meanings irrespective of facts and policies in contemporary context. It does not seem necessary to belabor the inadequacy of this conception of the process of interpretation and it may suffice to suggest that individuals of one age who work to control posterity by misplaced faith in the omnipotence of words of infinite abstraction are frequently to be disappointed.[14]

McDougal and Feliciano illustrate their point by noting that in some parts of the world, a bullet wound is termed "lead poisoning." They conclude that "the principle of restrictive interpretation must of course impose a limit upon such expansive extrapolations."[15]

The dispute between Schwarzenberger and McDougal and Feliciano on the restrictiveness of treaty interpretation is broadly reflected in legal literature. Further complicating the application of either interpretation in the present case is the mode of substance release—the fact that destruction of facilities would result in contamination of the environment—as contrasted to the delivery methods customarily associated with the use of poisons: ground-to-ground munitions, including grenades, shells, rockets, and missile warheads; air-to-ground munitions, including large bombs, dispensers, spray tanks, and rockets; and emplaced munitions, including generators upwind from a target, and mines.[16] Explicit codification is required to resolve the matter.

Applicability of the Principles of Humanity and Military Necessity. International law attempts to reconcile military necessity with humanitarian considerations. The 1899 and 1907 Hague conventions and the 1977 Geneva Protocol prohibit arms, projectiles, or materials or means of warfare calculated to cause "unnecessary suffering" or "superfluous injury."[17] From such statutory prohibitions and from practice, jurists have derived principles of military necessity and humanity. The principle of humanity forbids the use of weapons that are inherently cruel and offend minimum moral sensibilities.[18] The principle of military necessity allows for "such destruction and only such destruction as is necessary, relevant, and proportionate to the prompt realization of legitimate objectives."[19]

Reconciliation of these principles is difficult in theory as well as in practice. Tucker makes one attempt:

> The principle of necessity does not allow the employment of force unnecessary or superfluous to the purposes of war. Nor does the principle of humanity oppose human suffering or physical destruction. It is the unnecessary infliction of human suffering and the wanton destruction of property that is opposed by the principle of military necessity and the principle of humanity.[20]

Although a good conceptualization, Tucker's effort is not entirely satisfactory. It suffers, as do the definitions in the preceding paragraph, from a failure to define terms operationally. Without an empirical context, concepts such as weapons that "cause unnecessary suffering," that are "inherently cruel," and that offend "minimum moral sensibilities"; and destruction that is "necessary," "relevant," and "proportionate" are subjective. Interpreters may read into them what they find convenient.[21]

What is necessary, relevant, and proportionate for the attainment of military objectives may involve indiscriminate and morally offensive weapons and tactics. During World War II, the Allies and the Axis powers considered civilians as legitimate targets, since by so doing, they could destroy the morale of the enemy. In January 1943, the Casablanca Conference defined the primary purpose of the air war as "the progressive destruction and dislocation of German military, industrial, and economic system, and the undermining of the morale of the German people to the point where the capacity for armed resistance is fatally weakened."[22] This rationale resulted in massive bombing raids against cities, culminating in the dropping of the atomic bomb on Japan.

Judgments by postwar international tribunals on the legitimacy of civilian bombardment are ambiguous. The Nuremberg courts appear to have condoned such tactics by failing to raise the issue of German V-1 and V-2 rockets.[23] However, in the Shimoda case (1963), in which five residents of Hiroshima and Nagasaki sought compensation from the Japanese government for damages resulting from the atom bomb, the district court of Tokyo ruled that the American act violated international law. It stated that bombardment of an undefended city with a weapon that by its nature caused unnecessary suffering was illegal.[24]

Also at issue is the legitimacy of indiscriminate weapons designed to deter war. American decision makers consider the threat of nuclear attack a necessary and humane policy of deterrence. But to most of the other members of the United Nations, nuclear weapons are "contrary to the rules of international law and to the laws of humanity."[25] The United States in turn claims that this judgment is without "any legal basis."[26] The U.N. Charter offers little guidance on the matter. It denies members the right to threaten force (article 2, paragraph 4) but recognizes the "inherent right of individual or collective self-defense if an armed attack occurs . . . until the Security Council has taken the measures necessary to maintain international peace and security" (article 51). Goodrich and Hambro, in their extensive commentary on the charter, note the difficulty in reconciling the two articles in the nuclear age: "The development of atomic and hydrogen bombs and methods of delivery, creating the possibility that the initial armed attack will be decisive, make it highly unlikely that states will wait for such an attack to occur before exercising the right of self-defense.''[27]

Attempts to reconcile the humanistic intent of the law of war with the realities of practice have led some legal scholars to pessimistic views about the efficacy of this aspect of international law. McDougal and Feliciano, drawing on the thoughts of other scholars, contend:

> Historically the community of nations has never succeeded in outlawing any weapon which was of substantial net military utility. Weapon parity of course in particular situations induces reciprocal abstinence, but in general only weapons which were of marginal or indecisive military value and obsolete, or which were not deemed vital to the military establishments of one of the great powers, have been successfully prohibited.[28]

Scholars of like mind cite a number of historical examples from the period between 1900 and the end of World War II. On several occasions, specific prohibitions incorporated in the Hague conventions were ignored. Poison gas was used during World War I, as were mines intended to intercept commercial shipping. The prohibition against the dumdum bullet was effective only because steel bullets were accurate. Some agreements negotiated during the interwar period were also disregarded. The London Naval Treaty of 1930 and the London Procés-Verbal on Rules of Submarine Warfare, which required submarines to abide by the same rules as other warships, were ignored during World War II. Similarly the Hague prohibitions against indiscriminate bombardment of civilian targets were further elaborated in 1923 in the Hague Rules of Aerial Warfare. Although not adopted, they did provide a standard for behavior. It was not observed.[29]

Nevertheless belligerence during and particularly since World War II has at times followed the principles of the law of war. Most notable is the almost universal disuse of poison gas.[30] Indeed at the outbreak of hostilities in World War II, the British and French issued a declaration affirming their fidelity to the Geneva Protcol on Gas, except for purposes of retaliation. Their effort to elicit similar German assurances proved successful.[31] Practice during World War II, coupled with the interim pronouncements of statesmen and national leaders, have led two scholars to conclude that a "binding customary norm prohibiting at least the first use of the lethal or severely injurious types of chemical agents" has come into being.[32] Although we cannot be overly sanguine about national governments' abiding by international treaties or customary practice in time of war on the basis of this one example (as this book goes to press there are news accounts that the Soviets have used gas in Afghanistan), it demonstrates that international law has affected constraint on wartime behavior where there is reciprocity of interest associated with compliance.

In 1977 an effort to reconcile the principles of humanity and military necessity and apply them to civil nuclear energy facilities was consummated in an additional protocol to the 1949 Geneva conventions.[33] Article 56, although ambiguous and contradictory, addresses the permissibility of attacks against "nuclear electrical generating stations" in the broader context of "installations containing dangerous forces," including dams and dikes. Paragraph 1 declares that such works and installations "shall not be made the object of attack, even where these objects are military objectives, if such attack may cause the release of dangerous forces and consequent severe losses among the civilian population." The prohibition also extends to military objectives located in the vicinity of these works "if such attack may cause the release of dangerous forces and consequent severe losses among the civilian population." Stipulations that the losses must be severe for the prohibition to be applicable raise the question of what constitutes severity. This point is further complicated by the fact that irradiation might not result in death until years after exposure. Thus rather than being a clear prohibition, the paragraph's ambiguity may conceivably be used to justify an attack on a nuclear energy station.

Paragraph 2 further diminishes the strength of the prohibition:

The special protection against attack provided by paragraph 1 shall cease:

(a) for a dam or dyke only if it is used for other than its normal function and in regular, significant and direct support of military operations and if such attack is the only feasible way to terminate such support;

(b) for a nuclear electrical generating station only if it provides electrical power in regular, significant and direct support of military operations and if such attack is the only feasible way to terminate such support;

(c) for other military objectives located at or in the vicinity of these works of installations only if they are used in regular, significant and direct support of military operations and if such attack is the only feasible way to terminate such support.[34]

In effect, its inclusion allows an adversary to decide whether a nuclear facility provides "regular, significant, and direct support of military operations." Such a rationalization can usually be found. Paragraph 3 appears to attempt to minimize the implications of this exceptional clause by stipulating, "If the protection ceases . . . all practical precautions shall be taken to avoid the release of the dangerous forces." This qualification is not an adequate safeguard.

In addition, article 56 contains straightforward statements supplementary to the codification in the first two paragraphs. Paragraph 4 extends the prohibition to reprisals. Paragraph 5 attempts to minimize accidental destruction; it calls upon protocol signatories to avoid locating military objectives in the vicinity of installations containing dangerous forces. To facilitate identification, paragraph 7 advocates that nuclear installations be marked by three bright orange circles placed on the same axis. It adds that the absence of such markings in no way relieves adversaries of their obligations under the article.

Articles 54 and 55 also bear on the destruction of nuclear facilities, but they only add to the inadequacies. Article 54 is entitled, "Protection of Objects Indispensable to the Survival of the Civilian Population." Paragraphs 1, 2, and 4 codify clear prohibitions, only to be undercut in paragraphs 3 and 5. Paragraph 1 forbids "starvation of civilians as a method of warfare." Paragraph 2 stipulates:

> It is prohibited to attack, destroy, remove or render useless objects indispensable to the survival of the civilian population, such as foodstuffs, agricultural, drinking water installations and supplies and irrigation works for the specific purpose of denying them for their sustenance value to the civilian population or to the adverse Party, whatever the motive, whether in order to starve out civilians, to cause them to move away, or for any other motive.[35]

Paragraph 4 precludes reprisals. However, paragraph 3 specifies that the prohibition shall not apply to sustenance solely for members of an adversary's armed forces or objects directly supporting military operations as long as civilian populations are provided with sufficient food and water to prevent starvation or movements from their domiciles. Here, as in article 56, ambiguity is significant. What criteria determine "sustenance," "direct support of military action," or "inadequate food or water" for civilians? Combatants can usually find justification for using these exemptive clauses.

A second caveat provides an even more explicit basis condoning the destruction of nuclear facilities. Paragraph 5 stipulates:

> In recognition of the vital requirements of any Party to the conflict in the defense of its national territory against invasion, derogation from the prohibitions contained in paragraph 2 may be made by a Party to the conflict within such territory under its own control where required by imperative military necessity.[36]

In effect, this codification allows a state, in defense of its territory, to follow a scorched-earth policy to deprive an invading adversary of foodstuffs. In so doing, belligerents may be able to rationalize the release of radionuclides to contaminate foodstuffs that might fall into an antagonist's possession.

Article 55, "Protection of the Natural Environment," attempts to minimize consequences. It mandates that care be taken in warfare to protect the natural environment "against widespread, long-term and severe damage" resulting from acts that are "intended or may be expected to cause such damage to the natural environment and thereby to prejudice the health or survival of the population." Although the most straightforward of the articles reviewed, it logically contradicts the exemptive clauses that are part of articles 54 and 56.[37]

Furthermore the Geneva Protocol's treatment of nuclear facilities can be criticized for a lack of comprehensiveness. As the protocol now stands, it specifically addresses only one segment of the nuclear fuel cycle: nuclear electrical generating stations. However, large quantities of radionuclides are located in other fuel cycle installations: nuclear spent fuel installations, nuclear reprocessing plants, nuclear waste storage facilities, and nuclear fuel fabrication facilities. If the prohibition is to be comprehensive, these facilities must be included. Still another fault lies in the protocol's failure to address the permissibility of threats to destroy nuclear facilities. If their destruction is prohibited, consistency requires that the threat of destruction be prohibited in the codification.

In addition to the Protocol Additional, another relevant agreement opened for signature in 1977 was the Convention in the Prohibition of Military or Any Other Hostile Use of Environmental Modification Techniques. Although it does not address specifically nuclear facility destruction, what it says about the legality of environmental modification is important. The treaty forbids signatories from undertaking or encouraging other nations or groups to undertake "military or any hostile use of environmental modification techniques having widespread, long-lasting or severe effects as the means of destruction, damage or injury to any other State Party." "Environmental modification techniques" include but are not limited to deliberate manipulation of natural processes involving the earth's biota, lithosphere, hydrosphere, and atmosphere or of outer space resulting in such events as earthquakes, tsunamis, disruption of the ecological balance, or changes in ocean currents, the ozone layer, and the ionosphere. "Widespread" encompasses an area of several hundred square kilometers," "long-lasting," a period of months or approximately a season; and "severe," "serious or significant disruption or harm to human life, natural and economic resources or other assets."[38]

This convention fails to address nuclear facilities specifically. By drawing attention in article 2 to "deliberate manipulation of natural processes" and omitting nuclear facility specificity, it is uncertain whether the rationale for destroying nuclear installations for purposes other than contamination is prohibited. In addition, the treaty does not address the legitimacy of threats to manipulate the environment. Finally, the applicability of the convention is undermined by the Protocol Additional subsequently negotiated.

Conclusion. Given the principles discussed, is it permissible under current international law to destroy nuclear energy facilities deliberately? In view of the ambiguities of statute, custom, and interpretation, it is possible to argue either way.

From the perspective of military necessity, the release of radionuclides—whether intentional or coincidental to efforts to reduce an adversary's energy production—can be rationalized as consistent with bombing policies during World War II, with the present threat to use atomic weapons, with the ambiguities in the environmental modification convention, and with provisions of the 1977 Geneva Protocol. The destruction of facilities in proportionate reprisal would certainly be consistent with international practice. The fact that such an act would result in indiscriminate death and injury could be justified by arguing that war today is waged between entire nations, not simply between their military establishments.[39] It also can be argued that if the actual or threatened destruction would shorten a conflict and thus reduce suffering, the effort could be judged to be humanitarian.

The principle of humanity provides a different interpretation. An act of war is considered inhumane because of "the needlessness, the superfluity of harm, the gross imbalance between the military result and the incidental injury."[40] The deliberate release of radioactive products constitutes a different order of weapon from incendiary and fragmentation bombs, poison gas, or forces contained by dams and dikes. It is not simply the fact that radionuclides cause indiscriminate death and injury that is at issue in judging their legality. Rather it is the fact that they produce delayed, pernicious, somatic, and genetic effects that will not affect the outcome of a conflict. The destruction of nuclear energy facilities therefore should be prohibited.

In sum, international law is not sufficiently crystallized. The remainder of this chapter provides needed definition through formulating and rationalizing a draft treaty prohibiting the destruction of nuclear energy facilities for military purposes.

Related Proposals

In recent years, several proposals on radiological weapons, ecocide, and no-first-use of nuclear weapons have been made.

Radiological Weapons. Apprehension over radiological weapons is not new, but serious concern about them has been lacking until recently. The problem was raised in 1969 by Ambassador Arvid Pardo of Malta at the twenty-fourth session of the U.N. General Assembly. In an address, Pardo asserted that the issue deserved inquiry by the principal negotiating forum sponsored by the United Nations, the Conference of the Committee on Disarmament (CCD). Two matters concerned him. The first was "dirty" atom bombs—those that relied on radiation to kill and maim. The second was "the stockpiling and use of radioactive agents independently of nuclear explosions." Pardo argued that these agents could be derived from by-products of nuclear reactors "and could be used tactically or strategically—for instance in the form of radioactive dust or pellets—to contaminate a given area."[41] Despite the superpowers' reservations, Pardo persuaded the General Assembly to mandate the CCD to consider "effective methods of control against the use of radiological methods of warfare conducted independently of nuclear warheads," as well as "the need for effective methods of control of nuclear weapons that maximize radioactive effects."[42]

In 1970 the CCD took up the matter. After some discussion, the problem was dismissed as a theoretical one without "any practical significance."[43] The committee drew on a study submitted by the Dutch delegation, which argued that nuclear weapons maximized their potential by dealing a decisive blow against an opponent through the short-term lethal effects of blast, heat, and radiation created by the explosion rather than through the long-term effects of radiation. An increase in fallout, which would result in casualties weeks, months, or years after an attack, would serve little purpose from the military point of view. This concept would also hold true of long-lived radioactive agents produced without explosions. The study dismissed the idea of prohibiting the use of highly radioactive isotopes because of the difficulty of transporting them to a target area.[44]

The issue of radiological warfare lay dormant until 1976 when the United States and the Soviet Union held private bilateral discussions in Geneva on the matter. They followed a Soviet call in the 1975 U.N. General Assembly for prohibition of the development and manufacture of new types of weapons of mass destruction and of new systems of such weapons that the United Nations agreed should be considered by the CCD.[45] In the 1976 session of the General Assembly, Fred Iklé provided an idea of what the superpowers were considering when he noted that "rapidly accumulating radioactive materials have the potential for use in radiological weapons." Any strongly radioactive isotope, such as plutonium, could be dispersed through a conventional weapon, thereby contaminating substantial areas "for tens of thousands of years."[46] Iklé therefore proposed that the CCD consider radioactive materials as radiological weapons and prohibit their use. Making the matter an issue for multilateral concern in 1977 and 1978 was not pushed very hard as the superpowers continued their bilateral discussions.

In July 1979 at the signing of a new strategic arms limitation treaty in Vienna—SALT II—the Russians and Americans also announced agreement on a joint proposal to be submitted to a reconstituted CCD, the Conference on Disarmament.[47] The proposal stipulated that each state party to the treaty "undertakes not to develop, produce, stockpile, otherwise acquire or possess or use radiological weapons," defined as:

1. Any device, including any weapon or equipment, other than a nuclear explosive device, specifically designed to employ radioactive material by disseminating it to cause destruction, damage or injury by means of the radiation produced by the decay of such materials.

2. Any radioactive material, other than that produced by a nuclear explosive device, specifically designed for employment, by its dissemination, to cause destruction, damage or injury by means of the radiation produced by the decay of such material.[48]

The proposal also called upon each state party not to disseminate any radioactive material not defined as a radiological weapon that causes injury, damage, or destruction by means of the radiation produced by the decay of such material. At the same time it affirmed the right of nations to develop nuclear power for peaceful purposes.

An Ecocide Convention. Indirectly related to radionuclide destruction is Richard Falk's proposal for an international convention on the crime of ecocide. Falk's concern that the environment might be "selected as a 'military' target" grew out of American recourse in Vietnam to tactics intended to deny the enemy "the cover, the food, and the life-support of the countryside." The United States used herbicides, Roman plows, bulldozers, and weather modification. Anxious over the pernicious consequences of these acts, Falk called upon the international community to "take steps to strengthen and clarify international law with respect to the prohibition of weapons and tactics that inflict environmental damage, and designate as a distinct crime those cumulative war effects that do not merely disrupt, but substantially or even irreversibly destroy a distinct ecosystem."[49] Falk's proposed convention would define ecocide as

any of the following acts committed with the intent to disrupt or destroy, in whole or in part, a human ecosystem:

(a) The use of weapons of mass destruction, whether nuclear, bacteriological, chemical, or other;
(b) The use of chemical herbicides to defoliate and deforest natural forests for military purposes;
(c) The use of bombs and artillery in such quantity, density, or size as to impair the quality of the soil or to enhance the prospect of diseases dangerous to human beings, animals, or crops;
(d) The use of bulldozing equipment to destroy large tracts of forest or cropland for military purposes;
(e) The use of techniques designed to increase or decrease rainfall or otherwise modify weather as a weapon of war;
(f) The forcible removal of human beings or animals from their habitual places of habitation to expedite the pursuit of military or industrial objectives.[50]

Persons responsible for commitment, conspiracy to commit, incitement to commit, attempts to commit, and complicity in the commission of ecocide would be punishable. Later articles in the convention discuss liability for violation.[51]

No-First-Use of Nuclear Weapons. A third proposal suggests limiting the use of nuclear weapons to retaliation against an adversary who has used them first. Two groups of proponents of no-first-use are distinguishable. One group is concerned with gaining military advantage or reducing military disadvantage. The Chinese have persistently supported this policy since they exploded their first atom bomb. The reason may lie in their vulnerability to preemptive attack. The Soviet Union, reflecting its inferiority to the United States during most of the post-World War II period, has advocated a similar policy. However, the Soviet Union's emphasis has been on the prohibition of nuclear weapons, that is, a no-use policy. The United States, by contrast, has been of two minds on the question. Military doctrine as applied to its strategic forces calls for the use of nuclear weapons as second-strike weapons only, and the United States has unilaterally adopted a no-first-use strategy. American forces in Europe, however, are not prohibited from being the first to use nuclear weapons in retaliation for conventional attack. The policy for Europe reflects NATO's inferiority in conventional weapons compared to the Soviet Union.[52]

The second group of no-first-use proponents is not concerned with questions of military advantage but with the propriety of nuclear weapons in general.[53] They argue, "We have never accepted the idea that nuclear weapons, like poison gas or biological warfare, are weapons not to be used save in retaliation against their first use by an adversary."[54] They recognize that it is unlikely that nuclear weapons will be eliminated from the arsenals of major military powers. Therefore they propose to change the conditions that would result in the use of nuclear weapons. Because of the catastrophic consequences of crossing the nuclear threshold, they argue that the threshold should be raised by means of no-first-use agreements. Formal bilateral or multilateral agreements or unilateral declarations would establish an unwritten code of conduct among decision makers upon which expectations could be based and military tactics formulated; no-first-use commitments would be respected because of the threat of retaliation. To compensate for the impact this policy would have on military balances, conventional forces would have to be strengthened.

Conclusions. All three proposals are at least indirectly relevant. If the radiological weapons prohibition contemplated by the United States and the Soviet Union was consummated, a precedent for banning radionuclides in war would be established. However, unless nuclear installations were explicitly designated as weapons—this does not appear to be a component of discussions—they would not be covered given the varied rationale for their destruction.

Falk's ecocide convention is more comprehensive and explicit in its prohibition than is the 1977 convention on environmental modification. Although nuclear facilities are not mentioned, presumably the proposal could be elaborated to include this problem. However, there is another difficulty with this proposal. As it now stands, the draft treaty is so inclusive that if it were negotiated as a package, as Falk suggests, it might be impossible to obtain agreement beyond the generalities already found in the environmental modification treaty and the Geneva Protocol Additional. It can be argued that starting with an inclusive package is a good bargaining ploy, enabling the negotiators to reach the widest possible agreement.[55] However, the time consumed negotiating on the basis of an inclusive convention might be better spent if the most pressing or negotiable issues were each addressed independently, lest a nebulous prohibition be achieved. This point might have been a consideration for the United States and the Soviet Union to negotiate the question of radiological weapons outside the 1975 framework calling for a prohibition of new weapons of mass destruction.

Proposals prohibiting the first use of nuclear weapons treat a distinct problem. Nevertheless, the no-first-use principle conceivably could be applied to prohibit the threat to destroy or the actual destruction of nuclear energy facilities. However, such a prohibition would not be as unambiguous as one that declared these modes of coercion to be in violation of the law of war under any circumstances. An unambiguous statement is required to minimize destruction in war.

Utility of a Nuclear Energy Weapons Treaty

Since nations apply statutory and customary international law unevenly, it is useful to ask what purpose is served by prohibiting the wartime destruction of nuclear energy facilities. How great is the danger that such a treaty would simply codify illusory objectives? Several considerations indicate that such a treaty, or at least the exercise of negotiating it, would be worthwhile.

The novelty of the issue would make negotiation in itself educational for the participants and for the international community generally, whether or not it were successful. A consummated treaty would provide a common standard in an area that would otherwise depend on prudential judgment. Consequently a new element has its partisans within and outside the bureaucracy, and these partisans act as a pressure group working to assure its observance.[56] Furthermore, according to Abram Chayes,

The very promulgation of a formal prohibitory rule, validated by the political processes of the state, works to unify bureaucratic views, settle old arguments, and foreclose options. "An administrative mechanism," said Henry Kissinger, "has a bias in favor of the status quo, however arrived at." Once the treaty goes into effect, all the classical defects of bureaucracy become virtues from the point of view of arms control. Rigidity; absence of imagination, initiative, or creativity; unwillingness to take risks; operation by the book—all are enlisted in aid of compliance with the agreement.[57]

We cannot be overly sanguine about the effectiveness of education and bureaucratic inertia in restraining acts motivated by military necessity. But in the case of nuclear energy facilities, a treaty might be effective because the destruction of such facilities would result in indiscriminate damage that persists long after the conflict is over. Furthermore the legal status of an act of war or weapons is important in this case. Falk cogently argues that

> history records numerous efforts to proscribe the use of certain weapons that were considered at the time of their development to be especially destructive. These efforts largely failed, although the attempts to ban gas and germ warfare have enjoyed at least limited success. A study of the reasons why such prohibitions have been disregarded is instructive. The usual explanation—a reference to the primary of "military necessity"—is too abstract to identify specific pressures or to explain notable exceptions. It is doubtful that the United States would have introduced atomic bombs into World War II under the claim of military necessity if their status as weapons had been previously declared, with some formality, to be illegitimate. Would we, for instance, have been willing to attain an equivalent shortening of World War II (assuming the same quantum of damage) by the use of poison gas against Hiroshima and Nagasaki? I ask this question to suggest that the status of a weapon does appear to have some bearing upon the decision to use it.[58]

Both Chayes's and Falk's propositions have empirical support. Frederick Brown, in his study of restraints on the use of chemical weapons during World War II, finds that gas warfare has not assimilated into the military planning of any belligerent, in part because of legal prohibition. The only exception was Japan's use of gas to a limited extent in China. Brown concludes that on the basis of this experience, legal restraints appear to reinforce other existing restraints: "Treaty prohibition, though imperfect, reinforced both public and military dislike and fear of chemical warfare and provided a ready excuse for lack of substantive preparation."[59]

The key aspect of the treaty proposed here is the comprehensiveness of the prohibition against the release of radioactive products contained in nuclear energy facilities. These facilities include nuclear fuel fabrication, power, reprocessing, and waste-storage installations. The agreement also prohibits threats to destroy these works and reprisals. To minimize the possibility of accidental destruction, the location of military objectives within the vicinity of the facilities is forbidden.

Consultation among parties to the treaty and within the framework of the United Nations, including the Security Council, is stipulated to resolve problems in relation to the objectives and application of the agreement. To facilitate identification of the protected objects, the parties are urged to mark nuclear installations with three bright orange circles placed on the same axis, as specified in article 16 of annex I of the 1977 Geneva Protocol. The duration of the accord is unlimited; it may be reviewed and amended five years after its effective date. Nothing in the treaty affects the right of states to develop nuclear energy programs. Finally the signatories commit themselves to continue efforts to prohibit weapons of mass destruction and acts of war that violate the earth's ecology.

The treaty could be negotiated bilaterally or multilaterally, or its commitments could be declared unilaterally. The United Nations Conference on Disarmament, which began deliberations in 1979, would be the logical forum for multilateral negotiation. If states can agree that the destruction of nuclear facilities in time of war cannot be justified by military necessity, the proposed treaty should emerge relatively quickly as a new convention in international law.

Because of the importance of language in defining obligations, the draft of the proposed treaty is presented below in its entirety.

**A Proposed International Treaty on the
Prohibition of the Destruction for Military
Purposes of Nuclear Fuel Cycle Facilities
Containing Radionuclides**

The States Parties to this Treaty,

Determined to prevent needless suffering to persons exposed to radionuclides and their offspring, resulting from the destruction of nuclear fuel cycle facilities containing radioactive products;

Recognizing the important significance of international law that prohibits inhumane and indiscriminate methods of warfare including the June 17, 1925, Protcol for Prohibition of the Use in War of Asphyxiating, Poisonous and Other Gases, and of Bacteriological Methods of Warfare, and the April 10, 1971, Convention on the Prohibition of the Development, Production, and Stockpiling Bacteriological (Biological) and Toxin Weapons and on their Destruction, and conscious of the contribution which said Protocol and Convention has already made, and continues to make, to mitigating the horrors of war;

Recalling the stipulations of the 1899 and 1907 Hague Conventions on Land Warfare, and the 1977 Protocol Additional to the Geneva Conventions of 12 August, 1949, prohibiting the use of inhumane and indiscriminate weapons; and the 1977 Convention on the Prohibition of Military or Any or Hostile Use of Environmental Modification Techniques;

Recognizing that an agreement prohibiting the destruction for military purposes of nuclear fuel cycle facilities containing radionuclides is only one measure in an effort that must continue to ban all modes of combat that violate the earth's ecology and result in indiscriminate and inhumane suffering;

Convinced that the destruction for military purposes of nuclear fuel cycle facilities containing radionuclides would be repugnant to the conscience of mankind and that no effort should be spared to minimize this risk;

Have agreed as follows:

Article I

Each State Party to the Convention agrees that the release or threat to release radionuclides contained in nuclear fuel cycle facilities is prohibited under international law.

Article II

For the purposes of this treaty, nuclear fuel cycle facilities containing radionuclides include but are not limited to nuclear fuel fabrication facilities, nuclear power plants, nuclear spent fuel facilities, nuclear reprocessing facilities, and nuclear waste storage facilities.

Article III

The States Parties shall not locate any military objectives in the vicinity of the nuclear energy facilities defined by Article II.

Article IV

It is prohibited to make any of the works mentioned in Article II the object of reprisals.

Article V

The States Parties undertake to consult one another and to cooperate in solving any problems that may arise in relation to the objective of, or in the application of the provisions of, the Treaty. Consultation and cooperation pursuant to this article may also be undertaken through appropriate international procedures within the framework of the United Nations, including the Security Council which may review doubts and problems related to the treaty and take appropriate action in accordance with the Charter.[60]

Article VI

In order to facilitate the identification of nuclear fuel cycle facilities specified in Article II, the States Parties may mark them with a special sign consisting of a group of three bright orange circles placed on the same axis, as specified in Article 16 of Annex I to the 1977 Protocol Additional to the Geneva Conventions of 12 August 1949, and relating to the Protection of Victims of International Armed Conflict (Protocol I) stipulated in the Annex to this Treaty. The absence of such marking in no way relieves any State Party to a conflict of its obligations under the Treaty.[61]

Article VII

Nothing in the Treaty shall be interpreted as affecting the inalienable right of all the States Parties to develop research, production, and uses of nuclear energy for peaceful purposes.[62]

Article VIII

Any State Party may propse amendments to the Treaty. Amendments shall enter into force for each State Party accepting the amendments upon their acceptance by a majority of the States Parties to the Treaty and thereafter for each remaining State Party on the date of acceptance by it.[63]

Article IX

Each State Party to this Treaty affirms the recognized objective of effective prohibition of acts of war that result in mass destruction and ecocide, and undertakes to continue negotiations to prohibit such acts.

Article X

Five years after the entry into force of this Treaty, or earlier if it is requested by a majority of Parties to the Treaty by submitting a proposal to this effect to the Depositary Governments a conference of States Parties to the Treaty shall be held in Geneva, Switzerland, to review the operations of the Treaty, with a view to assuring that the purposes of the preamble and the provisions of the Treaty are being realized. Such a review shall take into account any new scientific and technological developments relevant to the Treaty. The review conference shall determine, in accordance with the views of a majority of those Parties attending, whether and when an additional review conference shall be convened.[64]

Article XI

The Treaty shall be of unlimited duration.

Article XII

(1) This Treaty shall be open to all States for signature. Any State that does not sign the Treaty before its entry into force in accordance with paragraph 3 of this article may accede to it at any time.

(2) This Treaty shall be subject to ratification by signatory Parties. Instruments of ratification and instruments of accession shall be deposited with the Governments of the Union of Soviet Socialist Republics, the United Kingdom of Great Britain and Northern Ireland, and the United States of America, which are hereby designated the Depositary Governments.

(3) The Treaty shall enter into force after the deposit of instruments of ratification by twenty-two Governments, including the Governments designated as Depositaries of the Convention.

(4) For States whose instruments of ratification or accession are deposited subsequent to the entry into force of this Treaty, it shall enter into force on the date of the deposit of their instruments of ratification of accession.

(5) The Depositary Governments shall promptly inform all signatory and acceding States of the date of each signature, the date of deposit of each instrument of ratification or of accession, and the date of entry into force of the Treaty and of the receipt of other notices.

(6) This Treaty shall be registered by the Depositary Governments pursuant to Article 102 of the Charter of the United Nations.[65]

Article XIII

This Treaty, the Chinese, English, French, Russian, and Spanish texts of which are equally authentic, shall be deposited in the archives of the Depositary Governments. Duly certified copies of the Treaty shall be transmitted by the Depositary Governments to the Governments of the signatory and acceding States.

In witness whereof, the undersigned, duly authorized, have signed this Treaty.

Done in triplicate, at   , this   , date of
.[66]

Annex: Works and Installations Containing
Dangerous Forces: International Special Sign[67]

1. The international special sign for works and installations containing dangerous forces, as provided for in Article VI of the Treaty, shall be a group of three bright orange circles of equal size placed on the same axis, the distance between each circle being one radius, as illustrated below.

2. The sign shall be as large as appropriate under the circumstances. When displayed over an extended surface it may be repeated as often as appropriate. It shall, whenever possible, be displayed on flat surfaces or on flags so as to be visible from as many directions and from as far away as possible.

3. On a flag, the distance between the outer limits of the sign on the adjacent sides of the flag shall be one radius of a circle. The flag shall be rectangular and shall have a white ground.

4. At night or when visibility is reduced, the sign may be lighted or illuminated. It may also be made of materials rendering it recognizable by technical means of detection.

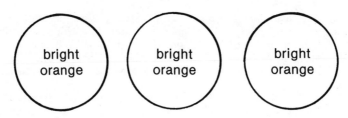

International Special Sign for Works and Installations Containing Dangerous Forces

Physical Options

Civil Defense

Although international law may provide some measure of restraint on state behavior, one cannot be overly sanguine about its effectiveness given its historic violation. As a result other alternatives must be available should it fail. Civil defense is one possibility. The purposes of civil defense are to prevent casualties, maintain public morale, ensure the operation of strategic industry, and limit property damage. Both active (military) and passive measures can contribute toward achieving these objectives.[68]

Active civil defense includes military interception and destruction of hostile forces before they can inflict damage. Front-line military forces are the primary defense, supplemented by point defenses specifically designed to protect potential targets. These point defenses typically include antiaircraft batteries, mine fields, artillery, and structures built of special materials. Although nuclear facilities have not been constructed with wartime bombardment in mind, their massive containment structures do provide some protection against attack. Additionally fences, alarms, cameras, and armed guards impede small groups of intruders bent on theft or sabotage. However, none of these prophylactic measures is sufficient against an attack with lethal weapons.

Although none of them is foolproof, some measures could improve point defenses. In time of crisis, military units could be stationed around installations to prevent assaults. Antiaircraft and artillery could be so situated to suppress bombardment. In the future, more sophisticated means of attack will require novel modes of defense. Homing missiles against cruise missiles and steel palings and tons of steel pellets lofted by explosives proposed to protect missile silos might be applicable to defense of nuclear energy facilities.[69]

Most nations also use passive civil defense to minimize casualties. These include education; life-supporting, well-stocked shelters; relocation strategies; and organization for rehabilitation. Many countries already have passive defense plans, but their comprehensiveness varies considerably.

The Swiss and Swedish programs are exemplary.[70] In Switzerland the law defines civil defense responsibilities shared by the federal government, cantons, communes, businesses, and individuals. A civil defense director supports the mayor at the basic commune level with a staff trained in public utilities; engineering services; fire rescue; medical services; welfare services; radiological, biological, and chemical services; intelligence and observation; industrial and institutional defense; and emergency operating center staff procedure. The most impressive aspect of the Swiss program is its shelters, which by the 1980s will be able to accommodate the entire population against most effects of nuclear weapons (or facility releases). This program is supplemented by underground hospitals and stores of government and military supplies.

Sweden's well-organized program is particularly notable for efforts to protect the economy by placing some industries, electrical power plants, and food stocks, as well as hospitals and command centers, underground. The Swedes also have considered the removal of radionuclides from reactors and other installations to deep underground sites during crises.[71] The population shelter program, not as extensive as that of the Swiss, centers on a relocation plan to remove 90 percent of the urban population to areas up to 250 mi. distant. Several successful relocation exercises have been carried out.

The programs of most other countries are comparatively undeveloped. For example, France does not have an urban shelter program. It relies on an evacuation plan premised on the belief that any nuclear conflict will be preceded by a lengthy period of crisis escalation. The Soviets maintain both an urban shelter program and an elaborate evacuation scheme, but critics believe that they appear more effective than they would be in the event of conflict. The civil defense programs of other countries fall within the Soviet-French spectrum.[72]

Given the efforts already underway in many countries to meet the threat of war, it would seem reasonable that consideration be given to the hazards posed by nuclear facility radiation. Precedent exists in the attention that some countries give toward meeting accidental releases. For example, in the United States the Nuclear Regulatory Commission requires licensees to develop emergency response plans to be coordinated with local and state agencies.[73] These plans are designed to afford timely warning and protection for nuclear employees and the general public. They should provide emergency control centers, means to monitor radionuclides, arrangements to quarantine contaminated areas and foodstuffs, assistance to the injured, and restoration of the plant. Preparedness should involve training and drilling of emergency personnel. Recent recommendations call for establishment of emergency planning zones up to 50 mi. from reactors.

Such measures should be elaborated to consider wartime threats. Populations should be educated about the problem and means to minimize it. Evacuation, the importance of which is underscored in Beyea's and U.S. government studies, should be planned, populations advised, and perhaps exercises undertaken. Radiation shelters should be provided for persons living close to nuclear installations. If shelters are not feasible, people should become acquainted with such expedient protective measures as remaining indoors and covering their nose and mouth with a cloth during the passage of the radioactive cloud. In addition, every household should have a supply of potassium iodine tablets. This stable iodine will block or dilute the intake of radioiodine if they are taken at the time of or a few hours before exposure.[74] Each household might also stock antinausea pills and breathing masks. Civil defense personnel should be equipped with special clothing and masks to operate in a radioactive environment.

Finally passive civil defense planning should include steps to deal with the long-term effects of radiation. These include diversion of radio-contaminated food, distribution of emergency food stocks, and decontamination procedures where feasible and necessary. Should these measures be undertaken in conjunction with active defense, both immediate and long-term casualties could be reduced significantly.

Facility Siting

Different modes of nuclear facility siting offer alternatives to diminish the consequences of contamination. Most nations already recognize that siting is important to protect populations from accidental releases. Therefore, reactors are commonly situated 15 to 20 mi. from urban areas, and support facilities are built in even more remote regions. Residential populations may be excluded entirely for several miles around facilities. Certainly one option to minimize exposure to contaminants is to extend these distances. However, this measure would entail increased costs, especially in the transmission of electricity from power plants and in finding acceptable locations. In small countries with high population density such as Israel or many nations in Western Europe, the problem is formidable. In other countries, including the Soviet Union, South Africa, Egypt, and Pakistan, all of which have remote regions, the option may be more attractive.

Remote siting need not be limited to land. Facilities can be placed on large lakes, inland seas, or the ocean. Power plants could be built on floating platforms surrounded by a breakwater, on floating vessels anchored to the marine floor, on artificial islands, or even undersea. In addition to isolating the reactors, so locating these plants or other facilities can provide utility companies greater siting flexibility, reduced manufacturing costs if facilities were standardized and constructed in shipyard type assembly areas, and lower aesthetic costs than does land siting. Furthermore should a meltdown occur, envelopment of material by surrounding water could help prevent material from becoming airborne. Indeed one of the positive lessons of the Three Mile Island accident was evidence that water has a significant capacity to retain radioactive iodine.[75]

There are also several potential problems with marine siting that must be weighed carefully: (1) higher transmission costs for reactors; (2) unique marine construction costs; (3) exposure to dangers peculiar to that environment, including ship collisions, fire, and explosive effects from shipping accidents, tsunamis, and sinking if the installations are located on platforms; (4) exposure to naval bombardment and, if off-site power is provided by onshore sites, the severing of cables providing electricity; and (5) contamination carried to and by sea breezes unless facilities are located great distances from shore.

Underground placement of facilities affords possibility for containing releases that other alternatives do not. Burying industry to avoid wartime destruction is not a new idea. During World War II, the Germans buried a number of installations to increase their resistance to bombardment. Among them were a 30,000 kW(e) public power plant at Mannheim and a private 8,000 kW(e) facility at Dentine. The Mannheim facility was a government experiment at the outset of the war to test the feasibility of the concept. The entire plant, including its single turbogenerator and boiler, was buried 50 ft. in the earth and protected with side walls 6 ft. thick and a roof of reinforced concrete 10 ft. thick. The installation's ability to resist attack was tested when a bomb landed 20 ft. from the outside walls. The generator sustained some damage that closed the facility for ten days. Then the plant was put back into operation, but with its capacity reduced to 22,000 kW(e).[76]

After Mannheim the Germans decided not to build additional power plants underground, apparently concluding that this alternative was not practical. Indeed later in the war when electrical capacity needed to be expanded at Mannheim, an old surface plant rather than the underground facility was enlarged. After World War II, the authors of the *United States Strategic Bombing Survey* assessed the German experience to learn how to protect plants in the future. It concluded that such precautions as brick or concrete walls around sensitive equipment, including circuit breakers, transformers, turbine generators, and personnel shelters, would "tend to reduce the extent of damage in case of attack."[77] However, like the Germans, the authors viewed the Mannheim experience skeptically and expressed doubt that any plant could be built economically to withstand attacks:

It can be assumed that in a plant designed especially for wartime operations, construction costs and operating costs are of secondary importance. In a peacetime plant the design strives to reach an economic balance between low initial cost and low operating cost. The controlling objectives in the design of a plant to meet wartime considerations differ so radically from those of a plant designed primarily for peacetime operations that it is impractical to combine the two objectives.[78]

Because of the inherent value of nuclear facilities, their large contribution to the economy, and the radiological threat posed by accident or war, countries today might find underground siting more attractive than they have in the past. Furthermore considerable progress has been made in underground construction since the *Strategic Bombing Survey* was published. Four small reactors have been placed below ground in Norway, Sweden, France, and Switzerland (table 4-1).[79] Underground siting may include the entire plant or the reactor alone. High-level liquid wastes, placed in underground containers, will be buried as solids in impermeable rock at depths of hundreds of feet. Other portions of the fuel cycle, including spent fuel and plutonium stores, also can be placed in underground bunkers. With limited processing of nuclear products at any one time, it would be unreasonable to place fuel fabrication or reprocessing installations below ground, assuming that preventing the release of products is the principal concern rather than loss of the plants.

With the exception of wastes buried many hundreds of feet down, nuclear installations would not be invulnerable to military acts, but they would be less vulnerable. External energy sources and coolant for reactors' spent fuel and high-level liquid wastes still could be disrupted by conventional bombardment, but the underground location would reduce significantly any radiological releases should a meltdown occur, assuming a large containment capacity for the cavern and rock and the maintenance of seals over surface penetrations. Should the seals fail, at least for reactors, the effect on the public might be similar to an equivalent release from a surface facility with containment isolation failure.[80] Radionuclides also would be released if an underground site were hit directly and cratered by a nuclear weapon. Even so, such siting would have some advantages. A single warhead would be unable to release the products of more than one installation in a multisite unit. If the nuclear weapon disrupted off-site power or coolant, burial would prevent fragmentation of the core and its additions to weapons fallout.[81]

Besides mitigating releases, underground siting affords several other advantages. Such facilities would be better able to withstand earthquakes than are their surface counterparts. The plants would be immune to other hazards, including storms, explosions, and aircraft crashes. Rock formations able to contain releases would obviate reinforced concrete contain-

Table 4-1
Underground Nuclear Reactors

Name and Location	Size	Purpose	Configuration		Depth (feet)	Reactor Chamber Dimensions (feet)
			Turbine Generator	Reactor		
Halden, Norway (BHWR)	25 MW(th)[a]	Experimental	None	Rock cavern	98	98' long 85' high 33' high
Agesta, Stockholm, Sweden (PHWR)	80 MW(th) 20 MW(e)	Heat and power	Above ground at reactor grade level	Rock cavern	49	88' long 66' high 54' high
Chooz, Ardennes, France	266 MW(e)	Power	Above ground	Rock cavern		138' long 146' high 69' wide
Lucerne, Switzerland	30 MW(th) 8.5 MW(e)	Experimental/ power	Rock cavern	Rock cavern		

Source: B.M. Watson, et al., *Underground Power Plant Siting* (Pasadena, Calif.: Environmental Quality Laboratory, California Institute of Technology, 1972), appendix, p. 1-2. Reprinted with permission.
[a]th: thermal (heat) output.

ment vessels. Because facilities would not protrude from the landscape, aesthetic values would be preserved, and the sites could be used for other purposes. Economies in transmission costs would be realized if power plants were situated near the centers of need, allowing waste heat to be put to industrial uses. Finally construction time would be shortened in the controlled environment of a cavern.[82]

Against such economies other costs must be weighed. The precise underground construction figures are unknown. According to one study addressing nuclear power plants, "Since there has never been a nuclear plant of 1,000 Mw(e) category built underground, it would be presumptuous to claim accuracy to any estimate."[83] Experience proves that prototype cost estimates are always low because of unforeseen problems. One study guessed that burial of a power plant could add as much as 40 percent to its cost.[84]

There are still other potential costs. Underground facilities are vulnerable to accidental flooding as a consequence of breaks in liquid coolant or heat transport lines. Groundwater could pose a similar problem. To minimize these dangers, facilities would need flood drainage systems and isolation valves for liquid conduits. Measures also would have to be taken to minimize the corrosive effects of groundwater seepage. Since underground installations are likely to be situated in tight quarters, inspection and maintenance would be more difficult than with surface sites.[85] Indicative of unforeseen problems are the releases of liquid wastes into the ground at the Hanford, Washington, disposal site.[86]

Many of these problems require technical solutions. They are not likely to be insuperable, and they may not even figure prominently in the cost-benefit debate. Rather this debate may focus on whether the additional security from burial is worth the investment. For nations with acute security threats, high population densities, limited territory, and few energy alternatives, such siting is worth considering.

Still another alternative would be to design surface and underground reactors and support facilities to flood from ground or river water if attacked. Conrad Chester argues that this would greatly reduce the area of contamination, virtually eliminating the aerosol threat. At the same time rivers are self-cleaning if the sediments are not disturbed.[87]

Improving Containment Effectiveness

Location of facilities in novel modes could address future plants, but it does not deal with the vulnerability of facilities now in operation or under construction. One alternative recently advanced by the Department of Engineering at the University of California, Los Angeles (UCLA), is a postaccident filter system (PAFS) for light water reactors.[88] The PAFS could either prevent containment failure caused by overpressurization resulting from releases of steam, noncondensible gases, or hydrogen burning or limit releases into the atmosphere if the containment was breached.

The UCLA design calls for a sand-gravel filter connected to the containment building by a duct 4 ft. in diameter with the flow of material facilitated by an exhaust fan. The filter, which could be adapted to reactors already in operation, would be situated in a reinforced concrete pit 120 ft. long, 100 ft. wide, and 24 ft. deep with additional backup high-efficiency particulate absorption (HEPA) and charcoal filters to remove radioiodine downstream. The system would retain 99 percent of the fission products, require minimal external services, be able to withstand missiles and fires arising within the power station and/or a large volume of moist air, not make access to other parts of the plant difficult, and be operable either from the main emergency control room or outside the control room. Cost estimates range from $1 million to $10 million. Figure 4-1 depicts one possible configuration. This proposal is not completely novel. Comparable arrangements have been applied to a British prototype 100 MW(e) British steam-generating heavy water reactor and a 300 MW(e) German prototype liquid metal breeder reactor.[89]

The design would not be effective against all accident sequences. For example, it could not contain the initial puff of radioactivity resulting from a steam explosion or a weapon-induced explosion. Still it promises to diminish the consequences of important accident sequences, perhaps by as great as a factor of ten in terms of fatal delayed cancers. Whether the system would be as effective in reducing wartime effects would depend on whether damage to the reactor was similar to accident scenarios wherein the PAFS is believed most effective and damage to the PAFS itself due to bombardment or sabotage.

Elevating Inherent Safety

Improving the inherent resistance of nuclear facilities and materials to release products is another way to diminish the consequences of military acts. Applicable theoretical, experimental, and practical attention has been devoted to efforts to diminish the effects of accidents. There has even been some attention paid to making facilities resistant to conventional weapons bombardment. In an unpublished paper, Theodore Taylor suggests criteria for a reactor that is resistant to such attacks. The reactor would use fuel coated with materials that effectively contain the afterheat of fission products without melting even if the pellets are broken. The fuel would have a large negative temperature coefficient; thus it would have the ability to

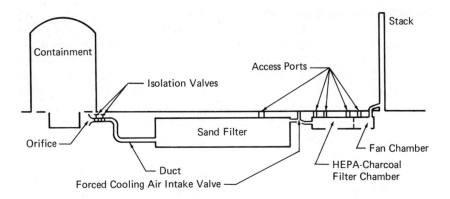

Source: David Okrent, et al., *Post-Accident Filtration as a Means of Improving Containment Effectiveness*, (Los Angeles: School of Engineering and Applied Science, December 1977), p. 3-22. Reprinted with permission.

Figure 4-1. General Arrangement of Postaccident Filtration System

decrease reactivity as temperature increased. It would have high heat capacity and allow for a large temperature difference between maximum fuel operating temperatures and temperatures fuel would begin to change. Core materials would not chemically react with any products with which they come in contact. Additional reinforcement against product releases would include a system to prevent control rods from falling out if jarred and sufficiently strong containment of materials in the pressure vessel.[90]

The concept of an inherently safe reactor is not mere speculation. Some are safer than others. Due to their high power density—the heat output per unit volume—fault conditions in LWRs produce a rapid temperature rise, allowing for a rapid meltdown. By contrast, temperatures will rise more slowly in reactors with lower power densities, such as the advanced gas and heavy-water-moderated designs. This slow rise allows time for the short-lived radionuclides to dissipate. Some people believe that the heavy water moderator in the Canadian Candu reactor could retain sufficient heat in the event of a loss of coolant to prevent any meltdown.[91] However, the design that best meets Taylor's criteria is the high-temperature gas reactor (HTGR). Its fuel is in the form of small particles of uranium and thorium carbide coated with several layers of carbon and a layer of silicon carbide to retain fission products.[92] Graphite is the moderator and helium, an inert gas that does not react with air, the coolant. With a negative temperature coefficient and the large heat capacity of helium, a loss of coolant in this design would result in a gradual rather than sudden release of radionuclides. Further adding to the design's safety is the location of the reactor core and the entire primary coolant system in a single, massive, reinforced concrete pressure vessel, which in turn is located in a containment building.

Many claims about the HTGR's safety derive from industry studies. Further independent analysis must be undertaken to verify them. The simulated accidents, more severe than those contemplated, should be reviewed, including the impact of a control rod ejection, and in the event of loss of coolant and containment rupture, igniting of the graphite moderator, as did occur in an accident at Windscale, England, in the 1960s, causing near total release of strontium and cerium fission products.[93] HTGRs would also have to overcome more obvious problems before they could be successfully marketed. One is cost; the design is more expensive than light water reactors.[94] The second is a 95 percent enriched uranium fuel that could be diverted to bomb making. Therefore fuel enriched as low as 6 percent is now being tested. If it proves practical, the weapons diversion problem would be eliminated.[95]

Inherent safety principles can also be applied to other aspects of the fuel cycle, notably high-level liquid wastes. Retention of these wastes in liquid form makes them particularly susceptible to releases if they are cut off from coolant. There is also the more acute problem of preventing leaks as a result of containment rupture, underscored by several such incidents due to corrosion.[96] Fortunately solidification into a glasslike or ceramic substance makes migration impossible except for water-induced leaching or nuclear explosion vaporization of products. The commercial reprocessing facility built in Barnwell, South Carolina, called for glassification after one year's cooling. This time could be reduced to perhaps three or four months if the liquid is converted into a granular state, allowing better heat dissipation.[97]

Alternative Energy Strategies

The most foolproof alternative to prevent nuclear facility destruction is cessation of reliance on this form of energy and adoption of other energy sources. Whether this alternative is viable depends on a number of factors, including technical feasibility, economics, and environmental, social, and political acceptability.[98] The literature contains considerable debate about whether alternatives can satisfy these criteria either in the short or long term.

Amory Lovins is the most articulate proponent of the position that they can. He argues that there are two mutually exclusive paths to energy growth, what he calls the hard and soft paths.[99] The hard path is currently favored by industrialized and industrializing countries. It relies on centralized energy production in the form of nuclear power, oil, natural gas, and coal. Presuming high consumption levels at a decreasing rate of growth, it proposes that coal and nuclear power bear the future energy burden as oil and gas poduction decline. Lovins criticizes this path on economic and social grounds. From the economic viewpoint, he finds a number of diseconomies of scale. Since centralized plants are limited in number, they cannot take advantage of the economies of mass production. Evidence suggests that the long lead times to build them result in increased cost escalation and interest charges (more than economies of scale decrease direct construction costs), making total capital cost per installed energy output more for very large than for small units. During this period, plants are exposed to changes in regulatory requirements, political moods, and technical evolution that often add to expenses. Utility cash flow may become a problem. Once the plants are built, they require distribution systems that often cost more than the plant itself. And they tend to be less reliable than smaller units, more expensive to repair, and needful of a large reserve capacity to provide power when they are not in operation. In addition to these diseconomies, large power plants exact larger environmental and socioeconomic costs than do smaller units. Beyond these problems lie their vulnerability to war, sabotage, accidents, and mistakes.[100]

In Lovins' view, these costs make the hard path unacceptable. In its place, he proposes an alternative soft path using the energy resources already available more efficiently by rapidly expanding the reliance on renewable energy resources afforded by the sun, wind, and vegetation. Energy efficiency can be achieved in a number of ways. Through better use of capital, design, management, and care, technical adaptations could result in more efficient engines and furnaces, less overlighting and overventilation in commercial buildings, recuperators for waste heat in industrial processes, and so on. Lovins contends, "There is overwhelming evidence that technical fixes are generally much cheaper than increasing energy supply, quicker, safer, of more lasting benefit."[101] He also believes they are more labor intensive and thereby increase employment. Furthermore they would not affect life-styles although social changes in the form of car pooling, smaller cars, mass transit, and recycling of materials would make a significant contribution to efficient use of energy.

Lovins acknowledges that technical fixes and social changes have their limits. To meet the foreseeable shortfall, he advocates the adoption of soft technologies: "a body of energy technologies that have certain specific features in common and that offer greater technical, economic and political attractions."[102] Such technologies are defined by five criteria:

They rely on renewable energy flows that are always there whether we use them or not, such as sun and wind and vegetation; on energy income, not on depletable energy capital.

They are diverse, so that energy supply is an aggregate of very many individually modest contributions, each designed for maximum effectiveness in particular circumstances.

They are flexible and relatively low-technology—which does not mean unsophisticated but rather, easy to understand without esoteric skills, acces-
sible rather than arcane.

They are matched in *scale* and in geographic distribution to end-use needs,
taking advantage of the free distribution of most natural energy flows.
[Thus such energy sources as centralized solar parks would not be considered soft technology.]

They are matched in energy quality to end-use needs.[103]

What distinguishes the hard and soft paths is not how much energy is used
but the technological and sociopolitical structure of the energy system.[104]

In Lovins's view, broad introduction of soft energy need not await
future technological development; economic technology is now available.
Solar collectors today can be used economically for the cooling and heating
of old and new buildings. Practical technologies are available to convert
agricultural and forestry wastes to methanol and other liquid and gaseous
fuels to the transport sector of the economy. Wind systems can produce
electricity, heat, pump, heat-pump, or compress air. Energy production at
the source of need rather than in central generating stations eliminates high
energy conversion and transmission costs. It does so simply, with low
replacement costs, slow obsolescence, high reliability, high volume, low
markup, and high employment. As for energy storage, often said to be a
potential problem, small units allow retention of low- and medium-
temperature heat at the point of use with water tanks, rock beds, or perhaps
fusible salts. In industry, wind-generated compressed air can be stored to
operate machinery, and organic conversion yields energy stored in liquid or
gaseous forms.

Lovins recognizes that soft energy cannot be introduced immediately.
Therefore he advocates fossil fuel transitional technologies coupled to
greater efficiencies. To conserve valuable oil and gas supplies essential for
petrochemical and other uses, he calls for interim adoption of sophisticated
clean coal technologies, available today, that could fill the transition period
with only a modest and temporary expansion of mining.

Lovins sees a number of comparative advantages in the soft path. In
contrast to high technologies whose success he contends is by no means
assured and whose environmental, social, and strategic costs are potentially
great, the soft path offers technical diversity, adaptability, and geographic
dispersion for a wide range of conditions for all regions of the world,
whether they be in highly industrialized countries or rural villages. In this
manner, energy can be provided without impinging upon the cultural patterns of a country and without dependence on a technical elite or commercial monopoly. The soft path offers an alternative to the nuclear energy
road to minimize malicious acts. Lovins stresses immediate implementation

of his proposals, arguing that delay pushes soft technologies so far into the future that there will be no credible fossil-fuel bridge to allow their implementation.

Lovins's thesis has elicited an enormous amount of controversy. In one congressional compendium alone, over two thousand pages of testimony is devoted to it.[105] In this collection, the U.S. Energy Research and Development Administration (ERDA), the principal predecessor of the Department of Energy, most cogently summed up the argument against it.[106] Agreeing with Lovins that there are already a wide variety of economical end-use and conservation technologies available, especially for buildings, that solar energy heating is likely to become increasingly attractive as the industry matures, and that cogenerating and district heating should receive increasing attention, ERDA took issue on two points. First, it rejected the contention that the hard and soft paths are mutually exclusive. Rather it suggested that there will be many circumstances where centrally supplied electrical power will be the most practical alternative because of its flexibility rather than because it is required as an energy form. Furthermore, hard energy will be necessary to make up the shortfall in oil and natural gas reserves that will occur whether energy growth remains constant or declines. ERDA anticipated a hybrid energy economy with both hard and soft technologies playing major roles. Whether soft energy could dominate would depend on technological development and economics. If it could not prove advantageous on these grounds, a national decision would have to be made that the noneconomic advantages were worth the extra cost.

A second point of contention concerns the capital investment required to implement the hard path. Lovins estimates $1 trillion. ERDA suggested half that amount consistent with the energy sector's historic share of fixed business investment btween 20 and 30 percent. As for the economies of soft technology, ERDA argued that these remained to be proved. Its computations suggested that soft technology is likely to be significantly more expensive than are hard technologies in meeting energy demands comprehensively. For example, although solar energy might economically meet 40 to 60 percent of space heating needs, sizing a system to meet 100 percent could escalate costs by an order of magnitude making solar energy much more expensive than hard technologies.

In sum, ERDA believed that Lovins did not establish that the soft path was either economically or technologically sound or that the hard path could not overcome technical and environmental problems. Because of the uncertainties, it concluded that abandonment of large-scale projects generally and nuclear energy specifically would be irresponsible.

The debate between Lovins and his opponents is not resolvable in this study. However, what is important from the wartime perspective is that Lovins's alternative may be worth considering by countries intent on diminishing, if not eliminating, the threat posed by nuclear energy during conflict.

Managing Alternatives through International and National Guidelines and Institutions

The alternatives presented thus far are not mutually exclusive, but their successful integration requires the commitment of all nations to the treaty prohibiting both the threat to destroy facilities and any efforts toward this end. Remaining alternatives would reinforce the treaty. Current hard and soft technologies would be exploited to their maximum. Nuclear energy would be utilized only if necessary. Inherently safe facilities with postaccident filtration systems would be located underground in remote regions protected by active defenses, while civil defense would be provided for populations downwind.

This scenario is not likely to be applied easily. Certainly it is not totally applicable to the hundreds of installations already in existence. Their vulnerability can be minimized only by the treaty, active defenses, and perhaps the adaption of postaccident filters. None is foolproof. Although the treaty may make a significant contribution toward eliminating facilities as legitimate targets, the prohibition may not be observed universally, and the filter or defenses against bombardment are not likely to be impervious. Only future construction can apply most of the remaining alternatives, and the advantages of these must be weighed against significant costs.

Because location and type of nuclear facilities bear on international security, expanded use of international as well as national regulatory bodies should be undertaken to ensure that new construction does not add to instabilities. To lay the empirical foundation for several alternatives, the performance of eight such institutions will be examined below: three U.S. government organizations—the Nuclear Regulatory Commission (NRC), the Energy Research and Development Administration (ERDA), and the Arms Control and Disarmament Agency (ACDA)—and five international bodies—the European Atomic Energy Agency (EURATOM), the Consultative Committee (Cocom), the International Bank for Reconstruction and Development (World Bank), the nuclear suppliers group, and the International Atomic Energy Agency (IAEA).

Nuclear Regulatory Commission

American nuclear regulatory policy has rapidly evolved over the last several years. The 1974 Energy Reorganization Act divided the Atomic Energy

Commission into two agencies: the Nuclear Regulatory Commission responsible for regulating nuclear energy, including export licensing, and a promotional institution, the Energy Research and Development Administration. In 1976 an executive order refined export licensing criteria. In 1977 ERDA's responsibilities were transferred to the new Department of Energy. Congress passed the Nuclear Anti-Proliferation Act in 1978, which further defined American export policy. For illustrative purposes, the NRC's performance applied to nuclear exports between 1974 and 1976 will be examined.

According to the 1954 Atomic Energy Act, the foundation upon which all subsequent legislation has been constructed, before nuclear material or facilities could be licensed, recipients first had to negotiate an Agreement for Cooperation. The agreement was a statement of principles containing the terms, conditions, duration, nature, and scope of cooperation; American safeguard rights; the recipient's declaration that nuclear material and facilities would not be used for research or development of nuclear weapons or for any other military purposes; and stipulations establishing when the recipient could transfer material.[107] During the 1974-1976 period, the agreement was prepared by ERDA, reviewed by the Department of State and the Congressional Joint Committee on Atomic Energy, and approved by the president. The following major criteria were used: the consistency of a nuclear relationship with American legal and policy requirements and with other agreements; the reasonableness of the scope of the desired cooperation; the availability of comparable assistance from other countries; national security implications; and the recipient's status with respect to Nonproliferation Treaty (NPT) and International Atomic Energy Agency (IAEA) obligations.[108]

Once an agreement was concluded, consummation of specific transactions required an NRC license granted on the basis of assurances that the transaction was consistent with American security. To establish this consistency, the commission submitted license applications to the executive branch. The State Department, acting as lead agency, consulted with ERDA, the Department of Defense, the Arms Control and Disarmament Agency, and the Department of Commerce to arrive at an evaluation. The following prescribed criteria had to be considered:

(1) The purpose of the export, (2) whether the export is covered by an agreement for cooperation, (3) whether the importing country has accepted and implemented acceptable international safeguards, (4) the adequacy of the importing country's accounting and inspection procedures and physical security arrangements to deal with threats of diversion of significant quantities of nuclear materials, (5) the importing country's position on nonproliferation of nuclear weapons, and (6) the importing country's understanding with the United States regarding the prohibition of using U.S. supplied material in the development of nuclear explosives.[109]

The Department of State homogenized the findings in a report it submitted to the commission.

The confidentiality of the licensing procedure makes assessment difficult. Nonetheless, the available evidence suggests that during the 1974-1976 period, NRC approval of exports was a matter of course. The Government Accounting Office (GAO) found that between 1975 and January 1976, forty of forty-nine applications were accepted, and the remainder were likely to be approved because the NRC had never disapproved a license under these procedures.[110] The GAO presented evidence that the success rate reflected less the merits of applicants than NRC deferral to a higher authority. Summing up its findings, the GAO noted:

> NRC officials have stated that there would probably be few cases where the Commission's judgment, in issuing an export license would differ from that of the Executive Branch. Should there be a difference at the end of the export license review process after all exchanges between NRC and the Executive Branch, NRC officials believe that they have the final decision making responsibility on whether to issue a license. However, NRC believes that because most export license transactions fall within the framework of agreements for cooperation developed by the Executive Branch with Congressional review and because the President is responsible for conducting foreign policy, his views on national security and foreign policy should be given great weight by NRC in making its export licensing decisions.[111]

This experience suggests that regulatory bodies that subordinate themselves to a higher authority that does not have coincidental interests cannot fulfill their mandate.

Energy Research and Development Administration

In 1976 ERDA, in collaboration with the Department of State, the NRC, and the Export-Import Bank, published a statement assessing "the environmental, social, technological, national, security, foreign policy, and economic benefits and costs to the United States associated with a continuation of nuclear power export activities through the year 2000."[112] The report also examined the costs and benefits of alternatives to current American nuclear export policy. The statement was issued in accordance with procedures and guidelines established by the National Environmental Policy Act of 1969, which mandates that the federal government anticipate and propose alternatives to acts that degrade the environment. Its purpose was

> to assist government decision makers, industry, and the public in making informed judgments on the proper nature, scope, and direction of the United States nuclear export activities, now and in the future, and the appropriate conditions that should govern those activities.[113]

The statement is interesting because of the way it assessed American nuclear export policy and because it demonstrated the limitations of unilateral undertakings of this sort.

According to the authors, "Nuclear power export activities have yielded numerous and significant benefits to the U.S. in such areas as national security, energy trade and employment."[114] From the security and foreign policy viewpoint, nuclear exports provided American leverage over the global market's development consistent with the objectives of the NPT. It helped allies reduce their dependence on fossil fuels from unreliable producers and contributed to the prosperity of all countries, including developing states. Economic benefits for the United States included $1.5 billion per annum from facility and fuel exports and a strong domestic nuclear industry.[115]

While making this case, the statement did not deny actual and potential costs: environmental costs from mining, land use for fabrication facilities and waste disposal, and possible security costs should materials be diverted for weapons by foreign governments or subnational groups. The statement also admitted that sabotage of facilities and materials was conceivable.[116] However, the authors stressed that there were limitations on the United States' capacity to diminish these costs unilaterally. Physical environmental impacts abroad were beyond the analysts' resources and authority. Furthermore any such undertaking could

> create risks of international repercussions arising from claims of encroachment by the U.S. on other nations' sovereignty since decisions as to the acceptability of risks of health and safety of a nation's citizens and to the physical environment traditionally have been reserved to the responsible sovereign government.[117]

As remedy, the authors suggested an international assessment: "Such an international assessment would not be constrained by sovereignty problems and could prove to be a useful tool in solving the worldwide issues related to the nuclear option."[118]

U.S. Arms Control and Disarmament Agency

ACDA is responsible for two relevant assessments: it advises the Department of State on its views on security implications arising from nuclear exports as part of the NRC review process, and it assesses the arms control implications of new weapons systems.

Nuclear Export Review. Although the executive branch's nuclear export assessments are confidential, congressional testimony on the American decision to export reactors to Egypt and Israel gave a glimpse into the workings of the Arms Control and Disarmament Agency in 1974. According to ACDA director Fred Iklé, the agency asked three questions:

> (1) Will the country that is to receive nuclear technology from the United States be likely to acquire such technology from other supplier nations?
> (2) Will the prospective technology transfer permit us to add further protective measures to the safeguards ordinarily applied, and thus permit us to take a step forward in separating the peaceful atom from the atom of war?
> (3) Is the region to which nuclear technology is exported free from latent or actual hostilities?[119]

Iklé testified that in ACDA's judgment, sales to the Middle East were likely to occur whether or not the United States was involved. Furthermore the United States would be more responsible than other countries in the application of safeguards. Iklé granted the region was prone to conflict. "But," he argued, "there is a countervailing consideration: the transfer of these power reactors can help strengthen United States influence in the area and thus help this administration and future administrations continue to bring peace to that area."[120]

Subsequent testimony by nongovernment witnesses questioned the basis of Iklé's testimony. Mason Willrich and George Quester, two prominent students of proliferation, argued that American responsibility in nuclear trade was no greater than that of the Soviet Union, France, Canada, and West Germany. Furthermore the American action would probably accelerate the military nuclearization of the region rather than deflate it.[121] As for Iklé's argument that nuclear trade enhanced American influence, Congressman Benjamin Rosenthal argued that there were less risky ways this objective could be achieved.[122]

Two rationales can account for the inadequacy of Iklé's remarks: either he was rationalizing as best he could a decision over which he had no control, and/or the agency's assessment was not as thorough as it should have been. Both explanations illuminate problems posed by efforts to anticipate the security implications of nuclear exports. The first underscores the importance of authority. Without it, review agencies simply acquiesce to the views of others. The second suggests the importance of honesty, thoroughness, and initiative by those undertaking such assessments. Without these attributes, assessments will not fulfill their objectives.

Arms Control Review. In 1975 Congress amended the Arms Control and Disarmament Act to require the executive branch to assess the arms control implications of military appropriations requests.[123] Although the amendment did not deal with nuclear exports, its function is similar enough to alternatives I will propose so that it provides a relevant case study.

The amendment stipulated that any agency preparing a legislative or budgetary proposal involving nuclear weapons or their delivery vehicles, or military programs with a total program cost in excess of $250 million or an annual cost of $50 million, or any other program the director of ACDA believed had a significant arms control, disarmament, or negotiation effect, was to provide the director with detailed information. In a manner he or she "deems appropriate," the director was to assess and analyze the programs in terms of their impact on arms control and submit his or her recommendations to the National Security Council, the Office of Management and Budget, and the government agency proposing the program. Congress also had authorization to review the statements.[124]

The Congressional Research Services (CRS) assessed arms control impact statements submitted to the Congress in January 1977. It found the statements deficient in at least two regards. First, they were not comprehensive. Of more than one hundred potential programs that could have been reviewed, the Department of Defense reported on twenty-one and ERDA on five. The Department of Defense asserted that seventy-six programs unaccountably did not have arms control impacts. However, the CRS concluded that at least half of these did have such implications.[125]

More serious was the shallowness of the assessments. It is apparent that the defense agencies narrowly interpreted their statutory obligations. The Department of Defense considered whether weapons systems violated existing arms control agreements or those being negotiated and their impact on strategic stability. The question of verification arose in only a few instances. ERDA addressed safety, security, and improved command and control of weapons systems; it did not, but should have, addressed programs' reduction of wartime destruction, economic burdens, inhibition of armaments races, and impacts on perceptions of national will and the military balance.

Several rationales account for the inadequacies of the statements. First, the CRS found three terms in the arms control legislation ambiguous: "program," "arms control," and "impact." The ambiguity left considerable room for interpretation and discretion by departments conducting the assessment. Second, it is unlikely that the departments responsible for weapons production were in the best position to judge critically arms control impacts. A third problem was the Arms Control and Disarmament Agency's lack of initiative in fulfilling its obligations. In the words of the CRS report:

> It was expected when the legislation of the arms control impact statements was passed that the U.S. Arms Control and Disarmament Agency, which has an arms control mission, would have a major if not leading role in preparing the impact statements. There is no evidence that such a role has materialized.[126]

To remedy these inadequacies, the CRS advocated more initiative by ACDA and clarification of the term *impact* by requiring that agencies conducting the review answer a detailed set of questions.[127]

European Atomic Energy Agency

Euratom is the first of five international institutions to be examined that were mandated responsibility to assess the implications of nuclear exports. Under article 103 of the Euratom charter, members of the organization contemplating an export were obligated to submit prospective export agreements for review by the community's commission. The commission was to conform the agreement to the organization's purposes. If the agreement impeded these purposes, the exporter was to be so advised. Presumably this advice would stimulate the supplier to reconsider its transaction. If reconsideration was not forthcoming, the exporter could be enjoined by the commission not to conclude the export until objections were removed, or the exporter might be obliged to subscribe to a ruling of Euratom's Court of Justice.[128]

Euratom has never been able to fulfill this obligation. Because of its failure to integrate Western Europe's nuclear programs, it has never felt that it had authority to act against the wishes of its individual members. Consequently it has gone through the exercise of reviewing exports, but in so doing has always acted in a perfunctory role concerning transactions.[129]

Consultative Committee

In the late 1940s, the United States initiated what eventually became the most concerted international effort to monitor exports of all sorts, nuclear power included, for security reasons. It tried to prevent a deterioration of American technological superiority over the Soviet Union by controlling exports to the Soviet bloc. In 1949 an organization known as the Consultative Committee (Cocom) was established to coordinate the policies of the United States and six others nations (England, France, Italy, the Netherlands, Belgium, and Luxembourg). In subsequent years the organization enlarged to include Norway, Canada, Denmark, West Germany, Portugal, Japan, Greece, and Turkey. In 1952 a parallel organization, known as Chicom, was established to control trade to mainland China.[130]

Cocom divided commodities into three categories: embargoed products, quantitatively regulated exports, and items to be kept under surveillance. At the height of the embargo, 1952-1955, over half of all internationally traded goods were on the lists. Nuclear power was among the items totally embargoed. In later years the number of such items was reduced drastically, and by 1958 only 10 percent of internationally traded goods were on the lists. Nuclear power plants and materials were among the goods deleted.[131]

With the exception of monitoring computers and sophisticated military ordnance, Cocom was not very successful. It did not prevent the Soviet Union from acquiring at least one sample of most embargoed goods. Those that the Soviets were unable to obtain at all or to obtain only in small quantities, they produced themselves after a relatively short delay, or they developed substitutes. One of the ironies of the embargo is that it may have forced members of the Soviet bloc into greater economic reliance on Moscow.

The principal reason for the failure of the embargo was lack of support. From the vantage point of Western Europeans, restraints impeded their World War II recovery that depended on exports. Furthermore they resented American coercion, which included threats to cut off assistance to uncooperative countries. Consequently many countries allowed their firms to engage in trade secretly.[132]

The Cocom experience underscores the limits of such international efforts. It was not a notable success because America's allies had different economic stakes, and American leverage was insufficient to induce cooperation. The Soviets contributed to the failure by finding adequate substitutes for the products they could not import.

World Bank

The International Bank for Reconstruction and Development conducts the most successful review. It loans monies for development projects such as electrical utilities, transportation networks, and industrial growth. In at least one case, Italy, it financed a nuclear reactor. To obtain financial support, a country must agree to a project appraisal. The bank approaches this review seriously. According to one source, "The Bank's experience has taught it that in project appraisal nothing should be taken for granted and that healthy skepticism is a cardinal virtue."[133]

The appraisal reviews six different aspects of a project: economic, technical, managerial, organizational, commercial, and financial. Economic assessments include the contribution of the project to the economy as a whole and its priority, taking into consideration scarce capital, managerial talent, and available skilled labor. Technical assessments appraise the engineering of the project: the appropriateness of proposed methods and processes, adequacy of design, construction scheduling, and potential causes of delay. Managerial appraisals review the demonstrated competence of the state in other enterprises. Organizational reviews estimate the organizational requirements to bring the project to fruition, as well as operational requirements. The commercial analysis calculates whether adequate arrangements were made for purchasing construction materials and, after completion of the projects, whether the requirements for maintaining the project will be available. Finally the financial analysis determines the soundness of the enterprise proposing to undertake the project.[134]

On the basis of its review, the World Bank can refuse to support a project, recommend its delay until there are more propitious circumstances, recommend extensive modifications, or adopt it. Once a loan is granted, the bank closely supervises the project to ensure that finances are properly and efficiently used. This supervision not only leads to success of the project but in the process reassures creditors, thereby keeping borrowing costs as low as possible.

In the caution and care it takes to evaluate projects, the World Bank is given high marks by its creditors. Its effectiveness is underscored further by the fact that it has had few failures. This is not to say that the bank's performance has been perfect. The bank has been criticized for not paying sufficient attention to ecological costs or the long-term implications of ventures. But on the whole the bank is praised for doing what it is supposed to do.[135] This success appears in no small part due to its authority and its willingness to use it, its rigorous review criteria, and the monetary commodity with which it deals.

Nuclear Suppliers' Group

In 1974 the United States, Canada, France, West Germany, Japan, the United Kingdom, and the Soviet Union embarked on a series of secret meetings to review what suppliers of nuclear material and equipment could do to ensure the peaceful application of nuclear energy. The suppliers' group—subsequently enlarged to include Belgium, Czechoslovakia, East Germany, Italy, the Netherlands, Poland, Sweden, and Switzerland —negotiated several agreements establishing nuclear transfer guidelines. The most recent to this writing was consummated in 1977.[136] It stipulates in detail nuclear materials and equipment that should trigger application of IAEA safeguards designed to detect illicit removal of nuclear material and establishes criteria for nuclear transfers. The criteria require recipients to agree that exports not be put to uses that would result in any nuclear explosive device and that exporters and importers mutually agree upon levels of physical security, although the measures assented to would be the responsibility of the recipient only. It also stipulates that restraint be exercised in the transfer of "sensitive facilities, e.g. reprocessing plants and weapons usable material" and encourages importers to accept multinational reprocessing and enrichment centers rather than national plants. Finally the document mandates that nuclear transfers not take place unless the importer agrees that in case of retransfer, the recipient is bound by the assurances that the first transferee assumed and that transactions involving stipulated sensitive material be approved by the original supplier.

Shortly after the consummation of the agreement, the Stockholm International Peace Research Institute, an authoritative international arms control investigative center, criticized the accord for being "essentially a gentlemen's agreement" rather than a binding treaty.[137] In a later assessment it noted that both the Soviet Union and the United States violated the stipulations in 1977 (and 1978-1979) by providing India with heavy water and enriched uranium, respectively, without New Delhi's agreement to accept international safeguards over its entire nuclear industry, although it acknowledged that both countries pressed India to apply stricter safeguards.[138] Despite the Stockholm Institute's cynicism and these violations, the guidelines provide a standard of international behavior that at the very least has forced exporters to think about the strategic implications of their trade. Whether they have contributed to policy changes is uncertain. It still may be too early to dismiss their effectiveness.

International Atomic Energy Agency

The IAEA is responsible for fostering peaceful nuclear development worldwide. To ensure that this development remains peaceful, it has devised a number of measures, including a set of physical-security guidelines for nuclear facilities and materials. Consummated in 1975 and revised in 1977, the guidelines recognize that although physical security is a national responsibility, the international implications of nuclear terrorism make it an issue of international concern.[139] Thus the guidelines stipulate how states should protect nuclear facilities and materials in use, transit, and storage. The agency reinforces its recommendations through participation in the development of legal instruments of international cooperation, technical assistance in the form of advice and training, publication of practical guidebooks on comprehensive physical protection systems, and maintenance of an information bank in the agency's library.

Although the guidelines are purely advisory and are not enforceable, the agency reports that they have been favorably received and used in some states for guidance in the preparation of national regulations. They are also referred to in the suppliers' guidelines. Although not universally applied, still they represent a benchmark that has had sufficient success to warrant ongoing negotiations to formulate a formal international convention.

Lessons

A number of lessons can be drawn from these cases concerning the requirements for successful national and international institutions that would anticipate and reduce the vulnerability of nuclear facilities not only to acts of war but possibly national and subnational diversion of nuclear materials for weapons purposes, subnational sabotage, and accidents as well. The illustrations suggest the importance of common interests between and within nations as a foundation for success, but it is a foundation difficult to construct. Cooperation can be forced as it was in Cocom. However, Cocom demonstrated that conflicting interests ultimately undermine restrictive trade practices. In like manner, different bureaucratic and political interests undermined the review efforts of the NRC and ACDA.

Common interests are necessary but may not be sufficient requisites for success. Optimally an organization must also have authority and independence to perform its functions. However regulatory authority in a community of sovereign nations may be difficult to acquire. This is true even within a national setting where, as the experiences of the NRC and ACDA suggest, even statutory mandates may not be genuine. In only one example, that of the World Bank, does authority seem to have been exercised. But the World Bank is unique in the sense that it deals with money, a commodity readily understood and desired.

Still the cases suggest that authority can be enhanced and the regulatory institution's integrity better ensured by a clear mandate. ACDA's ambiguous mandate, which allowed agencies conducting arms control reviews to be less thorough than they should have been, underscores the importance of clarity. To be effective, the mandate must not be so broad that an institution is asked to perform the impossible but not so narrow—as ERDA's American-focused impact statement—that it cannot perform its functions. The Congressional Research Service's criticism of ACDA's arms control assessments for failing to ask enough of the right questions and the World Bank's standards suggest the importance of detailed review criteria. And the IAEA's physical security guidelines and those of the suppliers' group illustrate alternatives, albeit imperfect due to their reliance on voluntary compliance, that establish standards of behavior that may have an impact.

Given the reluctance of nations to surrender sovereignty to international institutions but at the same time the desirability of establishing a criterion that addresses wartime risks, guidelines that provide the foundation upon which more authoritative institutions could be built offer the most immediate remedy. As they have in other cases, either the IAEA or the suppliers' group could establish detailed standards for exporters, importers, and nations building their own nuclear installations. The nuclear security assessment that follows illustrates what the guidelines might encompass.

The Nuclear Security Assessment

A. The Contemplated Nuclear Product

The vulnerability of nuclear facilities and consequences from destruction varies depending upon the installation and the material involved. Thus the risks associated with each would be considered first.

B. Economics of the Nuclear Product

States should make sure that the nuclear endeavor makes economic sense. This portion of the assessment would sensitize them to important economic considerations.

1. Costs: What are the projected planning, construction, and operational costs of the nuclear item? Answers to this question form the basis of the remainder of the economic assessment.
2. Energy requirements: What are the states' immediate and long-term energy needs? Such estimates will help determine whether nuclear power is the best source of electrical generation. Economical nuclear facilities today produce at least 600 MW(e). For many developing nations, this quantity may be beyond their requirements and/or distribution capabilities.
3. The economics of nuclear power: Is nuclear power economical both in the short and long term compared with other energy sources? Do its costs justify the quantity of scarce resources required—capital, domestic and foreign managerial talent, and skilled labor? The World Bank uses this question to determine whether it should support an export.
4. Balance of payments: Should the nuclear item be imported, what is the balance of payments impact? Many countries today have balance-of-payments problems and are unable to repay their debts. It is important that nuclear imports do not exacerbate this problem seriously.
5. Managerial and organizational skills: Does the state have sufficient skills to build and operate nuclear facilities? If not, what plans does it have to acquire them? The World Bank asks these questions to establish a project's worthiness.
6. Availability of energy backup systems: Does the nation have additional generating capacity for periods during which a nuclear power plant might be off-line? Nuclear power plants do not operate year-round. There are periods when they must be shut down for maintenance and repair. Light water reactors must be shut down during refueling. During these periods a nation must have generating capacity from other sources.

C. Wartime Vulnerability

Different military threats confront many nations. Some regions, such as Latin America, have been characterized by little interstate conflict, while in other areas, notably the Middle East, war has been chronic. Military capabilities vary widely; some states have nuclear weapons with advanced delivery systems, others have sophisticated conventional weapons, and still others have relatively unsophisticated armaments. This section attempts to anticipate the wartime vulnerability of nuclear facilities in different regions and suggests alternative means of dealing with the problem.

1. What is the nation's proneness to war? The year 1945 might serve as a benchmark for some historical perspective to this question while considering the points that follow.
2. What were the causes of wartime involvement: colonial, territorial, material, ideological? A statistical listing of wartime involvement without consideration of its roots would distort future projections.
3. Have wars taken place in the nation's territory? Many recent conflicts have arisen far from the combatants' territories. The Korean and Vietnam wars are two examples. This question, like the preceding one, is designed to eliminate distortion arising from a statistical review of wartime involvement.
4. What issues might cause the nation's participation in a war involving its own territory? Only such involvement should be of concern.
5. How likely is it that wars will involve nuclear weapons? For regions such as Europe where the probability of nuclear warfare is great, consideration should be given to whether destruction of the energy installations would be significant enough to warrant additional protective measures for nuclear plants.
6. What are the military capabilities of the nation's potential adversaries? Nuclear installations protected by reinforced concrete shells are likely to be invulnerable to all but the most sophisticated and accurately delivered munitions.
7. How much violence and discrimination is likely to occur between adversaries during conflict involving the nation's territory? Recent wars in the Middle East and South Asia were characterized by discriminate bombardment. Principal targets were the antagonist's armed forces rather than civilian populations and industrial centers. This behavior, reinforced by a treaty banning the destruction of nuclear energy facilities, might minimize attacks against nuclear power plants.
8. Where are facilities to be situated in relation to domestic populations and those of neighboring countries? The answer to this question is important in determining the location of installations.
9. What is the status of the nation's civil defense? Casualties resulting from radiological releases can be reduced if civil defense is adequate.

10. What impact will wartime vulnerable facilities have on international stability?
11. What options can reduce wartime vulnerability and threats to international stability?
    a. Nuclear facilities should not be built in regions where war is likely.
    b. Nuclear plants should be located in geographically remote areas, including oceans and seas for nations bordering such large bodies of water.
    c. Underground siting of nuclear plants and radioactive products should be considered.
    d. Containment shells and emergency safety systems should be reinforced.
    e. Nuclear installations should be fitted with a postaccident filtration system.
    f. Facilities that are inherently resistant to radionuclide releases should be considered.
    g. Active and passive civil defense should be strengthened.
    h. Facilities should be located in the vicinity of populations or valued land if this enhances international stability.
    i. An economic assessment of each option should be conducted.

These guidelines are simply suggestive. Certainly something along these lines would be a minimum national and international regulatory bodies could undertake to address the wartime problem. However, efforts need not stop with guidelines; more authoritative alternatives should be explored as well. These could include a standing working group either within the IAEA or the suppliers' group composed of international civil service social scientists and economists charged with the responsibility of making a security assessment of wartime and possible other nuclear energy risks prior to construction or acquisition of facilities and materials. Their conclusions, which would be advisory, could be circulated privately or publicized to take advantage of the moral authority of public opinion to influence decisions. Finally the historic trend toward greater international institutional involvement in minimization of nuclear threats might warrant the creation of a more authoritative international body to anticipate and regulate nuclear risks. Such an organization might control nuclear exports or license indigenous construction to ensure that nations live up to their international responsibilities as stipulated by a new international convention or guidelines.

Whether more authoritative institutions beyond guidelines are either desirable or attainable given the reluctance of nations to surrender sovereignty to international bureaucracies that might not reflect their interests is a matter that can be resolved only in international negotiation. Formulation of guidelines alone is probably the course of least resistance and consistent with IAEA practice in several areas—physical security, design, and operation of reactors. Since time is of the essence in establishing some standards, their definition would seem to warrant precedence.

5
Is There Merit in Ignoring Destruction of Nuclear Energy Facilities in War?

One final alternative remains: continuation of the current policy, which ignores the vulnerability of nuclear energy installations to destruction in war. The option is not without its merits. This study has focused on some of the worst consequences of wartime destruction. By so doing the argument can be attacked for possibly exaggerating the problem. No attempt has been made to establish a statistical probability that such consequences will occur. Some may take issue with the thesis that lesser or even some major releases are relatively significant compared to the inevitable consequences resulting from other military actions.

From another perspective, critics may object to my singling out the nuclear industry as a wartime risk. Other manufacturers, such as the chemical industry, also use dangerous products. Oil- and coal-burning power plants are likely to be as attractive as targets as are nuclear facilities if the objective of an attack is to destroy energy production. If each industry were subject to a wartime critique and forced to take remedial steps, products could not be produced economically. Finally, it may be asked, if nuclear energy vulnerability is a serious problem, why have policy makers and public critics not devoted more attention to it? Perhaps this neglect alone suggests that the issue presented here is overblown.

Such arguments are worth considering, but I believe that they are a weak rationale to ignore the problem. This study has looked at the worst consequences because they will occur if facility vulnerability is exploited under propitious weather conditions and if prophylactic measures are not taken. Rough probabilities for such occurrences can be established with detailed information about the nature of a particular nuclear installation, including its contents, local weather, and the effectiveness of military ordnance that states have available to them. Facility vulnerability is likely to increase in the future as nations acquire more lethal weapons, assuming that plants are not better protected. While one can argue that the releases upon which I have focused are conservative, it can also be argued that they are not conservative enough. For example, the reactor scenarios postulated destruction of one reactor per site. However, reactors increasingly are clustered in groups of two, three, or four to a site. Should the contents of two or more large plants be released, as well as on-site spent fuel in cooling ponds, the consequences would be considerably greater than those postulated, as will releases from some other reactors with greater concentrations of toxic products than LWRs, notably the LMFBR. Resulting fatalities and injuries will vary considerably among sites, depending on population density and civil defense. Whether they are significant will depend on one's definition of the term. Certainly in terms of early fatalities, the threat posed by the plants is not comparable to the devastating effects of nuclear weapons. On the other hand, one hardly can define the problem in conventional terms. The problem posed by land contamination alone is not trivial. Because of the long-term effects, Great Britain's Royal Commission on Environmental Pollution distinguished the nuclear industry from all others.

Why has wartime vulnerability received so little attention compared to other nuclear energy issues, such as nuclear weapons proliferation, terrorist acquisition of fissile material, sabotage, accidents, and waste disposal? There is no clear answer. However, policy makers have taken considerable time to grasp the implications of many nuclear issues. To sum up the argument, nuclear energy installations are attractive targets in time of war because of their energy production, intrinsic value, stores of nuclear weapons grade materials, potential to contaminate large regions, and susceptibility to politico-military manipulation. Destruction of facilities, although difficult, is not impossible. Explosives can breach structures containing radionuclides; releases also can result from disruption of mechanisms responsible for the removal of heat generated by the reactor core, spent fuel, and high-level liquid wastes. Military infiltration of installations would facilitate the task. Conventional bombardment with recently developed precision-guided munitions carrying high-explosive shaped charges is conceivable. Studies clearly support the vulnerability of facilities to nuclear attack.

In the event of a radionuclide release, the extensiveness of contamination will depend on the volume and composition of vented material, the height at which it is released from its containment, weather, and civil defense measures taken to minimize human exposure. Conceivably thousands of square miles could be contaminated, resulting in radiation sickness, thyroid damage, early and late death, genetic defects in future generations, and psychological traumatization. In many regions tens of thousands of people could be affected by releases unless they were relocated.

Given the potential of nuclear facilities to contaminate large areas, their destruction or threatened destruction may have significant implications for regional stability. Adversaries could manipulate the installations for purposes of deterrence, coercive diplomacy, and military strategy in a manner similar to the use to which the superpowers put nuclear weapons. Given the asymmetry in current or planned nuclear development in several regions, facility vulnerability may increase instability, assuming the party with the installations is sensitive to the problem and believes it faces a credible threat. Even where nuclear development is symmetrical, acquisition of the means to destroy installations and adoption of a strategic doctrine to do so could undermine the basis of current stability. The long-term consequences of release of facility radiation add humanitarian grounds for objection to facility manipulation or use as weapons.

It is fortuitous that in some of the most unstable regions—the Middle East, Korea, and southern Africa—nuclear energy has not advanced far, while in these and other regions, including South Asia, the most lethal conventional weapons have yet to be introduced. These facts allow time for reflection about the strategic meaning of facility vulnerability and for consideration of options: a treaty to prohibit the release or threat to release radionuclides contained in nuclear fuel cycle facilities, alternatives to diminish facility vulnerability to attack and the consequences derived therefrom (civil defense, alternative siting modes, improvement of the resistance of installations to release products), alternative energy sources, and international export and national review controls to ensure that installations are built to comply with international security.

The temptation of wartime threats or actions against nuclear energy facilities in many regions of the world adds a significant dimension to the problem of maintaining international peace and minimizing the consequences of war. These conclusions suggest that the vulnerability of nuclear energy facilities to military actions should be included in nuclear energy risk calculations.

Appendix A

Table A-1
Some Man-Rem Dose-Effect Coefficients for Delayed Effects
(per million person rem)

Whole body effects

About 130	Cancer deaths
25-250	Persons with identifiable dominant genetic defects over an average of five generations following exposure
12.5	Noninheritable congenital defects
0-500	Total extraconstitutionally or degeneratively diseased persons over an average of ten generations following exposure
42	Spontaneous abortions

Lung dose effects

0.6-1.6	Cancer deaths per year for the period five to at least twenty-seven years following exposure

Thyroid dose effects
Children under ten

0.5-3.0[a]	Cancer cases per year from five years until at least thirty years following exposure. (times about 0.04 for mortality during this period)
11-52[a]	Cases of thyroid nodules per year from five years until at least thirty years following exposure

Adults (persons over ten)

0-5-3.0[a]	Cancer cases per year from five years until at least twenty-five years following exposure. (times about 0.15 for mortality)

Source: American Physical Society Study Group on Light Water Reactor Safety, *Review of Modern Physics* 47, supplement no. 1 (Summer 1975):A-2.
[a]The lower value would hold if iodine-131 were rem-for-rem 0.3 times as effective in producing thyroid cancer than x-rays.

Notes

Introduction

1. This fact is underscored by the failure of the most comprehensive analyses of nuclear matters even to mention the subject. Among these studies are: American Physical Society Study Group on Light-Water Reactor Safety, *Review of Modern Physics* 47, supplement no. 1 (Summer 1975); U.S. Congress, Office of Technology Assessment, *Nuclear Proliferation and Safeguards* (New York: Praeger Publishers, 1977); Nuclear Energy Policy Study Group, *Nuclear Power Issues and Choices* (Cambridge, Mass.: Ballinger, 1977); Union of Concerned Scientists, *The Risks of Nuclear Power Reactors: A Review of the NRC Reactor Safety Study WASH 1400 (NUREG-75/014)* (Cambridge, Mass.: Union of Concerned Scientists, 1977); U.S. Energy Research and Development Administration, *Final Environmental Statement: U.S. Nuclear Power Export Activities*, ERDA 1542 (Springfield, Va.: National Technical Information Service, 1976); U.S. Nuclear Regulatory Commnission, *Reactor Safety Study*, WASH 1400 (Springfield, Va.: National Technical Information Service, 1975); and Albert Wohlstetter et al., *Swords from Plowshares: The Military Potential of Civilian Nuclear Energy* (Chicago: University of Chicago Press, 1979).

The only analysts who have treated this matter extensively in the public literature are Conrad V. Chester and Rowena O. Chester of the Oak Ridge National Laboratory. However, their assessments have been limited to the vulnerability of atomic facilities to nuclear attack. See their "Civil Defense Implications of the Pressurized Water Reactor in a Thermonuclear Target Area," *Nuclear Applications and Technology* 9 (December 1970):786-795; "Civil Defense Implications of a LMFBR in a Thermonuclear Target Area," *Nuclear Technology*, 21 (March 1974):190-200; and "Civil Defense Implications of the U.S. Nuclear Power Industry During a Large Nuclear War in the Year 2000," *Nuclear Technology* 31 (December 1976):326-338.

Remarks that facilities may be vulnerable to conventional weapons bombardment are scattered in the literature. These include Chester L. Cooper, "Nuclear Hostages," *Foreign Policy* 32 (Fall 1978):125-135; Henry W. Kendall in U.S. Congress, House, Committee on Foreign Affairs, Subcommittees on International Organizations and Movements and on the Near East and South Asia, *Hearings on U.S. Foreign Policy and the Export of Nuclear Technology to the Middle East*, 92 Cong., 2d sess., 1974, pp. 196-197; H.W. Lewis et al., *Risk Assessment Review Group Report of the U.S. Nuclear Regulatory Commission*, NUREG/CR-0400 (Washington,

D.C.: U.S. Nuclear Regulatory Commission, 1978), p. 45; Bennett Ramberg, "Destruction of Nuclear Energy Facilities in War: A Proposal for Legal Restraint," *World Order Studies Occasional Paper No. 7* (Princeton: Center of International Studies, Princeton University, 1978); Royal Commission on Environmental Pollution, *Sixth Report: Nuclear Power and the Environment* (London: Her Majesty's Stationery Office, 1976), pp. 123-124; and Theodore B. Taylor, "Reactor Safety Considerations Related to Sabotage and Wartime Bombardment of Nuclear Power Plants" (unpublished manuscript, 1968).

2. Royal Commission on Environmental Pollution, *Sixth Report.*

3. U.S. Strategic Bombing Survey, *German Electric Utilities Industry Report* (Washington, D.C.: Government Printing Office, 1947); Sir Charles Webster and Nobel Frankland, *The Strategic Air Offensive Against Germany, 1939-1945* (London: Her Majesty's Stationery Office, 1961); Oleg Hoeffding, *German Air Attacks Against Industry and Railroads in Russia, 1941-1954*, RM 6206-PR (Santa Monica, Calif.: Rand Corporation, 1970); U.S. Strategic Bombing Survey, *The Effects of Strategic Bombing on Japan's War Economy* (Washington, D.C.: Government Printing Office, n.d.); Robert Jackson, *Air War over Korea* (London: Ian Allan Ltd., 1973), pp. 141-143; Walter G. Hermes, *Truce Tent and Fighting Front* (Washington, D.C.: Government Printing Office, 1966), pp. 319-324; New York Times, *The Pentagon Papers* (New York: Bantam Books, 1971), pp. 502-505; U.S. Department of Defense, *United States-Vietnam Relations* (Washington, D.C.: Government Printing Office, 1971); Insight Team of the London Sunday Times, *The Yom Kippur War* (Garden City, N.Y.: Doubleday), pp. 203-205.

4. For a review of environmental warfare, see Richard A. Falk, "Environmental Warfare and Ecocide: A Legal Perspective," in Richard A. Falk, ed., *The Vietnam War and International Law: The Concluding Phase* (Princeton: Princeton University Press, 1976), pp. 287-303; Stockholm International Peace Research Institute, *Ecological Consequences of the Second Indochina War* (Stockholm: Almqvist and Wikseel International, 1976); Stockholm International Peace Research Institute, *Weapons of Mass Destruction and the Environment* (London: Taylor and Francis Ltd., 1977); and Stockholm International Peace Research Institute, *SIPRI Yearbook 1978* (London: Taylor and Francis Ltd., 1978), p. 43-51.

5. Cooper, "Nuclear Hostages"; Taylor, "Reactor Safety Considerations"; and Ramberg, "Destruction."

Chapter 1

1. For studies that review the effects of radiation on health from which the following discussion draws, see Peter Alexander, *Atomic Radi-*

ation and Life (Baltimore: Penguin, 1965); Jan Beyea, *A Study of Some of the Consequences of Hypothetical Reactor Accidents at Barsebäck*, PU/CES 61 (Princeton, N.J.: Center for Environmental Studies, Princeton University, 1978); Nuclear Energy Policy Study Group, *Nuclear Power Issues and Choices* (Cambridge, Mass.: Ballinger, 1977), pp. 159-196; Samuel Glasstone, ed., *The Effects of Nuclear Weapons* (Washington, D.C.: Government Printing Office, 1957); Samuel Glasstone and Philip J. Dolan, *The Effects of Nuclear Weapons* (Washington, D.C.: U.S. Department of Defense and U.S. Department of Energy, 1977); Alison P. Casarett, *Radiation Biology* (Englewood Cliffs, N.J.: Prentice-Hall, 1968); National Academy of Sciences, *The Effects on Populations of Exposure to Low Levels of Ionizing Radiation* (Washington, D.C.: National Academy of Sciences, 1972); Walter C. Patterson, *Nuclear Power* (Baltimore: Penguin Books, 1976); Union of Concerned Scientists, *The Risks of Nuclear Power Reactors: A Review of the NRC Reactor Safety Study WASH 1400 (NUREG-75/014)* (Cambridge, Mass.: Union of Concerned Scientists, 1977); U.N. Scientific Committee on the Effects of Ionizing Radiation, *Sources and Effects of Ionizing Radiation* (New York: United Nations, 1977); U.S. Nuclear Regulatory Commission, *Reactor Safety Study*, WASH 1400 (Springfield, Va.: National Technical Information Service, 1975), appendix VI.

2. For elaboration see Alexander, *Atomic Radiation*, pp. 26-67; Nuclear Energy Policy Study Group, *Nuclear Power Issues*, pp. 160-162; Casarett, *Radiation Biology*, pp. 31-33; and Glasstone and Dolan, *Effects*, pp. 576-577.

3. Glasstone, ed., *Effects*, pp. 473-480.

4. U.S. Nuclear Regulatory Commission, *Reader Safety Study*, p. 73.

5. *Federal Register*, January 18, 1977, pp. 2558-2561.

6. Royal Commission on Environmental Pollution, *Sixth Report: Nuclear Power and the Environment* (London: Her Majesty's Sationery Office, 1976), pp. 123-124.

7. U.N. Scientific Committee, *Sources and Effects*, pp. 5-8.

8. Nuclear Energy Policy Study Group, *Nuclear Power Issues*, pp. 166-167.

9. R.O. Chester and F.O. Hoffman, "Protection of the Thyroid Gland in the Event of Releases of Radioiodine," *Nuclear Safety* 19 (November-December 1978):732-740; Nuclear Regulatory Commission, *Reactor Safety Study*, appendix VI, p. 9-26.

10. Robert A. Conrad et al., "Twenty-Year Review of Medical Findings in a Marshallese Population Accidentally Exposed to Radioactive Fallout," in U.S. Congress, House, Committee on International Relations, Subcommittee on International Security and Scientific Affairs, *Hearings on First Use of Nuclear Weapons: Preserving Responsible Control*, 94th Cong., 2d sess., 1976, pp. 195-198.

11. National Academy of Sciences, *Effects on Population*, p. 48.

12. Ibid., pp. 46-47.

13. Robert J. Lifton, *Death in Life* (New York: Random House, 1967).

14. Ibid., pp. 106-107.

15. Ibid., p. 107.

16. Ibid., p. 108.

17. Ibid., pp. 178-181.

18. Nuclear Regulatory Commission, *Reactor Safety Study*, appendix VI, pp. 1-2.

19. Studies that consider the impact of radiation on agriculture and livestock include David W. Bensen and Arnold H. Sparrow, eds., *Survival of Food Crops and Livestock in the Event of a Nuclear War* (Oak Ridge, Tenn.: U.S. Atomic Energy Commission Technical Information Center, 1971); Eric B. Fowler, ed., *Radioactive Fallout, Soils, Plants, Foods, Man* (Amsterdam: Elsevier, 1965); National Research Council, Committee to Study the Long-Term Worldwide Effects of Multiple Nuclear Weapons Detonations, *Long-Term Worldwide Effects of Multiple Nuclear-Weapons Detonations* (Washington, D.C.: National Academy of Sciences, 1975); R. Scott Russell, ed., *Radioactivity and Human Diet* (Oxford: Pergamon Press, 1966); Nuclear Regulatory Commission, *Reactor Safety Study*, appendix VI, pp. E-1-E-44; and U.N. Food and Agriculture Organization, *Agricultural and Public Health Aspects of Radioactive Contamination in Normal and Emergency Situations* (Rome: Food and Agriculture Organization of the United Nations, 1964).

Chapter 2

1. Donald McIsaac, ed., *United States Strategic Bombing Survey* (New York: Garland, 1976), 6:49.

2. Ibid., p. 3.

3. Oleg Hoeffding, *German Air Attacks Against Industry and Railroads in Russia, 1941-1945*, RM-6206-PR (Santa Monica, Calif.: Rand Corporation, 1970).

4. Robert Jackson, *Air War over Korea* (London: Ian Allan Ltd., 1973), pp. 141-143; Walter G. Mermes, *Truce Tent and Fighting Front* (Washington, D.C.: Government Printing Office, 1966), pp. 319-324; New York Times, *The Pentagon Papers* (New York: Bantam Books, 1971), pp. 502-505; U.S. Department of Defense, *United States-Vietnam Relations* (Washington, D.C.: Government Printing Office, 1971), vol. 6; Insight Team of the London Sunday Times, *The Yom Kippur War* (Garden City, N.Y.: Doubleday, 1974), pp. 203-205.

5. McIsaac, *Bombing Survey*, pp. 26-39.

6. Theodore B. Taylor, "Reactor Safety Considerations Related to Sabotage and Wartime Bombardment of Nuclear Power Plants" (unpublished manuscript, 1968), pp. 7-9.

7. Stockholm International Peace Institute, *Ecological Consequences of the Second Indochina War* (Stockholm: Almqvist and Wiksell International, 1976), pp. 49-63.

8. Taylor, "Reactor Safety Considerations," p. 7a.

9. For studies that elaborate the operation of light water reactors, see American Physical Society Study Group on Light Water Reactor Safety, *Review of Modern Physics* 47, supplement no. 1 (Summer 1975); Walter C. Patterson, *Nuclear Power* (Baltimore: Penguin Books, 1976); U.S. Atomic Energy Commission, *The Safety of Nuclear Power Reactors and Related Facilities*, WASH 1250 (July 1973); and Richard E. Webb, *The Accident Hazards of Nuclear Power Plants* (Amherst: University of Massachusetts Press, 1976).

10. "World List of Nuclear Power Plants," *Nuclear News* 22 (August 1979):69-87.

11. A. Birkhofer and D. Smidt, "Implications of WASH-1400 in the Federal Republic of Germany," *Transactions of the American Nuclear Society* 24 (1976):331-332, and John A. Richardson, "Summary Comparison of West European and U.S. Licensing Regulations for LWRs," *Nuclear Engineering International* 21 (February 1976):32-41.

12. Atomic Energy Commission, Division of Reactor Technology, *Soviet Power Reactors, 1970*, WASH 1175 (Springfield, Va.: National Technical Information Service, 1970); Joseph Lewin, "The Russian Approach to Nuclear Reactor Safety," *Nuclear Safety* 18 (July-August 1977):438-450; Philip R. Pryde, "Nuclear Energy in the Soviet Union," *Current History* 77 (October 1979):115-118; "Nuclear Parks Answer Soviet Siting Problems," *Nuclear Engineering International* 24 (December 1979); 5; N. Dollezhal and Y. Koryakin, "Nuclear Power Engineering in the Soviet Union," *Bulletin of the Atomic Scientists* 36 (January 1980):33-37.

13. Consultant Workshop, Sandia Laboratories, *Summary Report on Workshop on Sabotage Protection in Nuclear Power Plant Design*, SAND 76-0637 (Washington, D.C.: Nuclear Regulatory Commission, 1977), pp. 20-21.

14. Ibid., p. 17.

15. U.S. Nuclear Regulatory Commission, *Reactor Safety Study*, WASH 1400 (Springfield, Va.: National Technical Information Service, 1975), appendix VI, pp. 2-1-2-4.

16. D.H. Slade, ed., *Meteorology and Atomic Energy 1968*, TID-24190 (Oak Ridge, Tenn.: U.S. Atomic Energy Commission, 1968), p. 305.

17. Nuclear Regulatory Commission, *Reactor Safety Study*, appendix VI, p. 4-3.

18. Lynn T. Ritchie, et al., "Effects of Rainstorms and Runoff on Consequences of Atmospheric Releases From Nuclear Reactor Accidents," *Nuclear Safety* 19 (March-April 1978):220-238, and Jan Beyea, "Short-term Effects of Catastrophic Accidents on Communities Surrounding the Sundesert Nuclear Installation," *Testimony Before the Energy Resources Conservation and Development Commission of California*, mimeo. (December 3, 1976).

19. Geostrophic winds reflect the impact of the earth's rotation and atmospheric pressure. Below 600 to 900 m, the trajectory of these winds are deflected by the friction created by topography. According to one source, "Over a relatively smooth level surface, such as a prairie or a body of water, surface wind is deflected to an angle about 20-25 degrees left of the geostrophic direction (Northern Hemisphere) and may have a speed of 60-70% of the geostrophic wind. In contrast, surface winds over a rough surface may be turned as much as 45 degrees left of the isobar and reduced to one-third of the geostrophic wind speed." Arthur N. Strahler, *The Earth Sciences*, 2d ed. (New York: Harper and Row, 1971), p. 237. Local surface winds also will be affected by other factors. For example, convection circulation caused by unequal heating of the atmosphere results in sea, land, valley, and mountain breezes. For elaboration, see ibid., pp. 232-255, and Glen T. Trewartha, *An Introduction to Climate*, 4th ed. (New York: McGraw-Hill, 1968), pp. 65-119.

20. Nuclear Regulatory Commission, *Reactor Safety Study*, appendix VI, p. II-17, and Jan Beyea, *A Study of Some of the Consequences of Hypothetical Reactor Accidents at Barsebäck*, PU/CES 61 (Princeton: Center for Environmental Studies, Princeton University, 1978), p. II-16.

21. Jan Beyea, "In the Matter of Long Island Lighting Company (Jamesport Nuclear Power Station, Units 1 and 2)," *Direct Testimony of Dr. Jan Beyea*, Case No. 80003 (New York: New York State Board on Electric Generation Siting and the Environment, May 1977).

22. Private communication from Jan Beyea.

23. Nuclear Regulatory Commission, *Reactor Safety Study*, pp. 11-15.

24. Nuclear Energy Policy Study Group, *Nuclear Power Issues and Choices* (Cambridge, Mass.: Ballinger, 1977), p. 167.

25. Beyea, *A Study of Some of the Consequences*, pp. I-6, I-10.

26. Ibid.

27. Nuclear Regulatory Commission, *Reactor Safety Study*, Main Report, pp. 9, 11.

28. American Physical Society Study Group, *Review*, p. 72.

29. *Los Angeles Times*, July 23, 1978, p. 1.

30. C.V. Chester and R.O. Chester, "Civil Defense Implications of a Pressurized Water Reactor in a Thermonuclear Target Area," *Nuclear Applications and Technology* 9 (December 1970):787.

31. Congressional Research Service, *Nuclear Proliferation Factbook* (Washington, D.C.: Government Printing Office, 1977), p. 203.

32. American Physical Society Study Group, *Review*, p. S17.

33. A curie is a "unit of quantity of radioactive materials. Any amount of radioactive material decaying at the rate of 3.7×10^{10} disintegrations per second is 1 curie of that material." U.S. Atomic Energy Commission, *The Safety of Nuclear Power Reactors and Related Facilities*, WASH 1250 (Washington, D.C.: U.S. Atomic Energy Commission, July 1973), pp. 4-5.

34. U.S. Nuclear Regulatory Commission, Office of International Programs, *Proceedings from the NRC-PAEA Spent Fuel Storage Meeting* (Springfield, Va.: National Technical Information Service, 1978).

35. Atomic Energy Commission, *Safety*, pp. 1-59.

36. U.S. Atomic Energy Commission, *GESMO: Generic Environmental Statement on the Use of Recycled Plutonium in Mixed Oxide Fuel LWR's* (Washington, D.C.: U.S. Atomic Energy Commission, 1974), p. IV D-17, and Dean C. Kaul and Edward S. Sachs, *Adversary Action in the Nuclear Power Fuel Cycle: I. Reference Events and Their Consequences*, SAI-121-612-7803 (Schaumburg, Ill.: Science Applications, 1977), chap. A.

37. Kaul and Sachs, *Adversary Action*.

38. Atomic Energy Commission, *Safety of Nuclear Power Reactors*, pp. 4-77.

39. Kaul and Sachs, *Adversary Action*.

40. Michael Flood, "Nuclear Sabotage," *Bulletin of the Atomic Scientists* 32 (October 1976):36, n. 20. In a private communication Conrad Chester of the Oak Ridge National Laboratory took exception to this contention. He argues that "fuel elements are stored with appropriate neutron absorbers to prevent nuclear reactions. More importantly, LWR core designs are undermoderated, and compression will further reduce reactivity." Letter dated April 9, 1980.

41. Kaul and Sachs, *Adversary Actions*, table III-1.

42. Ibid., table III-2.

43. Jan Beyea, "The Effects of Releases to the Atmosphere of Radioactivity from Hypothetical Large-Scale Accidents at the Proposed Gorleben Waste Treatment Facility," unpublished paper (Princeton: Center for Environmental Studies, 1979).

44. Conrad V. Chester and Rowena O. Chester, "Civil Defense Implications of the U.S. Nuclear Power Industry During a Large Nuclear War in the Year 2000," *Nuclear Technology* 31 (December 1976):331.

45. Thomas A. Brown, "Missile Accuracy and Strategic Lethality," *Survival* 18 (March-April 1976):52.

46. Chester and Chester, "Civil Defense Implications of the Pressurized Water Reactor," p. 794.

47. Chester and Chester, "Civil Defense Implications of the U.S. Nuclear Power Industry," p. 329.

48. Tom Gervasi, *Arsenal of Democracy: American Weapons Available for Export* (New York: Grove Press, 1977), p. 170, and Cecil I. Hudson and Peter H. Haas, "New Technologies: The Prospects," in Johan J. Holst and Uwe Nerlich, *Beyond Nuclear Deterrence: New Aims* (New York: Crane, Russak, 1977), p. 128.

49. Hudson and Haas, "New Technologies," pp. 108-113.

50. Anne Hessing Cahn and Joseph J. Kruzel, "Arms Trade in the 1980s," in Anne Hessing Cahn, et al., *Controlling Future Arms Trade* (New York: McGraw-Hill, 1977), p. 61, and Stockholm International Peace Research Institute, *World Armaments and Disarmament, SIPRI Yearbook 1975* (Cambridge, Mass.: MIT Press, 1975), p. 332.

51. Bruce L. Welch in U.S. Congress, Joint Committee on Atomic Energy, *Hearings on Possible Modification or Extention of the Price-Anderson Insurance and Indemnity Act*, 93d Cong., 2d sess., 1974, p. 285.

52. Nuclear Energy Policy Study Group, *Nuclear Power, Issues and Choices* (Cambridge, Mass.: Ballinger, 1977).

53. For studies that elaborate the sabotage problem, see Comptroller General of the United States, *Security at Nuclear Power Plants—At Best Inadequate* (Washington, D.C.: General Accounting Office, 1977); Flood, "Nuclear Sabotage," pp. 29-36; Mitre Corporation, *The Threat to License Nuclear Facilities* (Washington, D.C.: Mitre Corporation, 1975); Joseph D. Schleimer, "The Day They Blew Up San Onofre," *Bulletin of the Atomic Scientists* 30 (October 1974):24-27; U.S. Congress, House, Committee on Interior and Insular Affairs, Subcommittee on Energy and the Environment, *Oversight Hearing on Nuclear Reactor Security Against Sabotage*, 95th Cong., 1st sess., 1977.

54. Beyea, *A Study of Some of the Consequences*, p. 103. For a critique of this conclusion see H.L. Gjorup, et al., *A Technical Evaluation of Jan Beyea's Report, "A Study of Some of the Consequences of Hypothetical Reactor Accidents at Barsebäck,"* Riso-M-2108 (Roskilde, Denmark: Riso National Laboratory, October 1978). See also a German study that suggests that the worst accident, including a core meltdown and containment failure, could result in 14,000 casualties within 20 km and 104,000 late cancers within thirty years in the German context: "West German Risk Report Echoes Rasmussen," *Nuclear Engineering International* 24 (September 1979):4.

Chapter 3

1. Chester L. Cooper, "Nuclear Hostages," *Foreign Policy* 32 (Fall 1978):125-135. Reprinted with permission from *Foreign Policy*. Copyright 1978 by the Carnegie Endowment for International Peace.

2. Ibid., p. 135.

3. These distinctions are drawn from Alexander L. George, et al., *The Limits of Coercive Diplomacy* (Boston: Little, Brown, 1971), pp. 16-32.

4. Thomas C. Schelling, *Arms and Influence* (New Haven: Yale University Press, 1966), p. 22.

5. The views of Secretary of Defense Harold Brown are pertinent in these regards: "U.S. central systems, of course, remain the ultimate deterrent, and are inextricably linked to the defense of Europe. Augmentation of NATO's long-range theater nuclear forces based in Europe, however, would complete the Alliance's continuum of deterrence and defense, and strengthen the linkage of U.S. strategic forces to the defense of Europe. Indeed, increased NATO options for restrained and controlled nuclear responses reduce the risk that the Soviets might perceive—however incorrectly—that because NATO lacked credible theater military responses, they could use or threaten to use their own long-range theater nuclear forces to advantage.

We have already developed the flexibility with our theater nuclear forces to execute:

limited nuclear options that permit the selective destruction of particular sets of fixed enemy military or industrial targets;

regional nuclear options that, as one example, could aim at destroying the leading elements of an attacking enemy force; and

theaterwide nuclear options that take under attack aircraft and missile bases, lines of communication, and troop concentrations in the follow-on echelons of an enemy attack."

Harold Brown, *Department of Defense Annual Report Fiscal Year 1981* (Washington, D.C.: Department of Defense, 1980), p. 94.

6. Schelling, *Arms*, p. 23.

7. Samuel Glasstone, ed., *The Effects of Nuclear Weapons* (Washington, D.C.: U.S. Government Printing Office, 1957), p. 471.

8. Yehezkel Dror, *Crazy States* (Lexington, Mass.: Lexington Books, D.C. Heath and Co., 1971), p. 5.

9. This might avoid what Robert Jervis refers to as the "security dilemma": acquisition of weapons for one's own security, which diminishes

the security of an opponent. See his "Cooperation under the Security Dilemma," *ACIS Working Paper No. 4* (Los Angeles: Center for Arms Control and International Security, University of California, Los Angeles, 1977).

10. Stockholm International Peace Research Institute, *World Armaments and Disarmament: SIPRI Yearbook 1978* (London: Taylor and Francis Ltd., 1978), p. 22.

11. For a review of Soviet civil defense, see Leon Gouré, *War Survival in Soviet Strategy: USSR Civil Defense* (Miami, Fla.: Center for Advanced International Studies, 1976).

12. Fred Kaplan, "The Soviet Civil Defense Myth," *Bulletin of the Atomic Scientists* 34 (March 1978):14-20, and Fred M. Kaplan, "Soviet Civil Defense: Some Myths in the Western Debate," *Survival* 20 (May-June 1978):113-120.

13. Zhores A. Medvedev, *Nuclear Disaster in the Urals* (New York: W.W. Norton, 1979).

14. This policy is referred to as "assured destruction." In his 1980 annual report to the Congress, Secretary of Defense Brown indicated that while assured destruction in the "bedrock of nuclear deterrence" the United States also retains the option "to respond at a level appropriate to the type of scale of a Soviet Attack." Harold Brown, *Department of Defense Annual Report Fiscal Year 1981* (Washington, D.C.: Department of Defense, 1980), pp. 65-66.

15. See Catherine Kelleher, *Germany and the Politics of Nuclear Weapons* (New York: Columbia University Press, 1975); Anne H. Cahn, "Determinants of the Nuclear Option: The Case of Iran," in Onkar Marwah and Ann Schulz, *Nuclear Proliferation and the Near-Nuclear Countries* (Cambridge, Mass.: Ballinger, 1975), pp. 185-204; Dimitrije Sesrinac Gedza, "Yugoslavia and Nuclear Weapons," *Survival* 18 (May/June 1976):116-117; Albert Wohlstetter, et al., *Swords from Plowshares: The Military Potential of Civilian Nuclear Energy* (Chicago: University of Chicago Press, 1979), pp. 111-125; and Roger Gale, "Nuclear Power and Japan: Proliferation Option," *Asian Survey* 18 (November 1978):1117-1133.

16. Lewis A. Dunn and Herman Kahn, *Trends in Nuclear Proliferation, 1978-1995*, HI-2336-RR/3 (Croton-on-Hudson, N.Y.: Hudson Institute, 1976).

17. For a list of these facilities, see "World List of Nuclear Power Plants," *Nuclear News* 22 69-87; B.M. Jasani, "Nuclear Fuel Fabrication Plants" and "Nuclear Fuel Reprocessing Plants" in Stockholm International Peace Research Institute, *Nuclear Proliferation Problems* (Cambridge: MIT Press, 1974), pp. 70-98; and Congressional Research Service, *Nuclear Proliferation Factbook* (Washington, D.C.: Government Printing Office, 1977), p. 203.

18. Harold Brown, *Department of Defense Annual Report Fiscal Year 1979* (Washington, D.C.: Department of Defense, 1978), pp. 68-69, and Graham H. Turbiville, "Invasion in Europe," *Army* 26 (November 1976):16-21.

19. Central Intelligence Agency, "East Germany-Poulation" (map) 501041 (Langley, Va.: Central Intelligence Agency, July 1973).

20. Discussions of the future of nuclear energy in the Middle East include U.S. Congress, House, Committee on Foreign Affairs, Subcommittees on International Organizations Movements on the Near East and South Asia, *Hearings on U.S. Foreign Policy and the Export of Nuclear Technology to the Middle East*, 92 Cong., 2d sess., 1977; *U.S. Senate Delegation Report on American Foreign Policy and Nonproliferation Interests in the Middle East: Report Pursuant to Senate Resolution 167 of May 10, 1977*, 95th Cong., 1st sess., 1977; Atomic Energy Establishment and Nuclear Power Plants Authority, Egypt, "Projected Role of Nuclear Power in Egypt and Problems Encountered in Implementation of the First Nuclear Plant," IAEA CN 36/574 (Vienna: International Atomic Energy Agency 1977); *Los Angeles Times*, August 25, 1979, pp. 1, 7; "Israel Looks Again at Nuclear," *Nuclear Engineering International* 25 (January 1980):6.

21. Fact Finding Group on Nuclear Power, *Report to the New Zealand Government* (Wellington, New Zealand: E.C. Keating Government Printer), p. 257.

22. Steven Rosen, "Nuclearization and Stability in the Middle East," in Marwah and Shulz, *Nuclear Proliferation*, pp. 176-178.

23. Gregory Francis Winn, "Arms, Attitudes and Decisions: The Probability of Conflict in Northeast Asia with Particular Emphasis on the Korean Peninsula" (Ph.D. diss., University of Southern California, 1978).

24. Young-Sun Ha, "Nuclearization of Small States and World Order: The Case of Korea," *Asian Survey* 18 (November 1978):1134-1151.

25. Ibid., p. 1136, n. 6.

26. *Los Angeles Times*, December 5, 1978, pt. III, p. 11; "Peking Plans to Cancel Reactors," *Nuclear Engineering International* 24 (August 1979):5; and *Los Angeles Times*, April 6, 1980, P. I, p. 6.

27. J..G. Bartholomew, et al., *Atlas of Meteorology* (Chicago: Denoyer-Geppert Co., 1899), plate 12.

28. For a review of relations, see William N. Brown, *The United States and India, Pakistan, and Bangladesh* (Cambridge, Harvard University Press, 1972), pp. 106-226.

29. Zalmay Khalizad, "Pakistan: The Making of a Nuclear Power," *Asian Survey* 16 (June 1976):580-592.

30. International Institute for Strategic Studies, *The Military Balance, 1978-1979* (London: International Institute for Strategic Studies, 1978), p. 66.

31. Bartholomew et al., *Atlas*.

32. *Nuclear News* 22 (September 1979):52.

33. Analyses of Iran's foreign policy from which this assessment draws include Rouhollah K. Ramazani, *Iran's Foreign Policy, 1941-1973* (Charlottesville: University of Virginia Press, 1975); Shahram Chubin and Sepehr Zabih, *The Foreign Relations of Iran: A Developing State in a Zone of Great-Power Conflict* (Berkeley: University of California Press, 1974); and Shahram Chubin, "Iran's Military Security in the 1980s," *Discussion Paper No. 73* (Los Angeles: California Seminar on Arms Control and Foreign Policy, 1977).

34. Randall Stokes, "External Liberation Movements," in Ian Robertson and Phillip Whitten, *Race and Politics in South Africa* (New Brunswick, N.J.: Transaction Books, 1978), pp. 203-219; Colin Legum, "Conclusion: Looking to the Future," in Gwendolen M. Carter and Patrick O. Meara, *Southern Africa in Crisis* (Bloomington, Ind.: Indiana University Press, 1977), pp. 258-267.

35. International Institute for Strategic Studies, *Military Balance*, pp. 49-50.

36. Barbara Rogers and Zdenek Cervenka, *The Nuclear Axis* (New York: New York Times Books, 1978), p. 193. In 1979 South Africa may have conducted a test in the atmosphere.

37. Ibid., pp. 172-173.

38. Bartholomew et al., *Atlas*, p. 12.

39. A list of U.S. nuclear power plants is found in "World List of Nuclear Power"; reprocessing installations in Congressional Research Service, *Facts on Nuclear Proliferation* (Washington, D.C.: Government Printing Office, 1975), pp. 84-85; and mixed oxide fuel fabrication facilities in Jasni, "Nuclear Fuel Fabrication," p. 84 to 85.

40. Conrad V. Chester and Rowena O. Chester, "Civil Defense Implications of the U.S. Nuclear Power Industry During a Large Nuclear War in the Year 2000," *Nuclear Technology* 31 (December 1976):337-338.

41. Meteorological Office (Naval Division), Air Ministry, *Weather in the Mediterranean* (London: H.M. Stationery Office, 1936), pp. 11 10-13, and 16.

42. See Fuad Jabbar, *Israel and Nuclear Weapons* (London: Chatto and Windus, 1971), pp. 133, 146-147; Shlomo Aronson, "Israel's Nuclear Options," *ACIS Working Paper No. 7* (Los Angeles: Center for Arms Control and International Security, University of California, 1977); Robert W. Tucker, "Israel and the United States: From Dependence to Nuclear Weapons," *Commentary* 110 (November 1975):29-43; and Rosen, "Nuclearization," pp. 157-184.

43. For studies that call attention to the dangers of nuclear weapons proliferation, see Robert J. Pranger and Dale R. Tahtinen, *Nuclear Threat*

in the Middle East (Washington, D.C.: American Enterprise Institute, 1975); Lewis A. Dunn and Herman Kahn, *Trends in Nuclear Proliferation, 1975-1995*, HI-2336 RR/3 (Croton-on-Hudson, N.Y.: Hudson Institute, 1976), pp. 114-147; Albert Wohlstetter, Henry Rowen, and Richard Brody, "Middle-East Instabilities and Distant Guarantors (and Disturbers) of the Peace: The Arab-Israeli Case," *Discussion Paper* (Los Angeles: California Seminar on Arms Control and Foreign Policy, 1978), pp. 25-34; Wohlstetter et al., *Swords from Plowshares*; William Epstein, *The Last Chance: Nuclear Proliferation and Arms Control* (New York: Free Press, 1976), pp. 98-109.

Chapter 4

1. Fred C. Iklé, "Can Nuclear Deterrence Last Out the Century?" *Foreign Affairs* 51 (January 1973):267-285.

2. Ibid., p. 281.

3. Richard A. Falk, "Environmental Warfare and Ecocide: A Legal Perspective," in Richard A. Falk, ed., *The Vietnam War and International Law: The Concluding Phase* (Princeton, N.J.: Princeton University Press, 1976), p. 290. My discussion of legal remedy is a revision of Bennett Ramberg, "Destruction of Nuclear Energy Facilities in War: A Proposal For Legal Restraint," *World Order Studies Program Occasional Paper No. 7* (Princeton, N.J.: Princeton University Center of International Studies, 1978).

4. Ann Van Wynen Thomas and A.G. Thomas, Jr., *Legal Limits on the Use of Chemical and Biological Weapons* (Dallas, Tex.: Southern Methodist University Press, 1970), p. 39.

5. *Statute of the International Court of Justice*, art. 38.

6. The distinction between "sources" and "subsidiary means" for determining the rules of law appears in ibid. See also Thomas and Thomas, *Legal Limits*, p. 227. The extent to which General Assembly resolutions are authoritative is controversial. Falk contends that "the record of reliance on such resolutions in areas of arms control, space and human rights . . . can, where intended by a large majority of governments, declare and create law." Falk, "Environmental Warfare," p. 289.

7. For texts, see Leon Friedman, ed., *The Law of War* (New York: Random House, 1972), pp. 204-397.

8. Ibid., pp. 229, 318.

9. Ibid., p. 249.

10. Ibid., pp. 424-456.

11. U.S. Arms Control and Disarmament Agency, *Arms Control and Disarmament Agreements: Texts and History of Negotiations* (Washington, D.C.: U.S. Arms Control and Disarmament Agency, 1975), pp. 118-124.

12. U.S. Arms Control and Disarmament Agency, *Documents on Disarmament, 1971* (Washington, D.C.: Government Printing Office, 1973), pp. 885-886, 910-913; U.S. Arms Control and Disarmament Agency, *Documents on Disarmament, 1966* (Washington, D.C.: Government Printing Office, 1967), pp. 789-799.

13. George Schwartzenberger, *The Legality of Nuclear Weapons* (London: Stevens and Sons Ltd., 1958), pp. 27-29. Julius Stone also holds this opinion. See his *Legal Control of International Conflict* (New York: Rinehart, 1954), p. 343.

14. Myres S. McDougal and Florentino P. Feliciano, *Law and Minimum World Public Order* (New Haven: Yale University Press, 1961), pp. 664-665.

14. Ibid., p. 665.

15. See Myres McDougal, Harold D. Lasswell, and James Miller, *The Interpretation of Agreements and World Public Order* (New Haven: Yale University Press, 1967).

16. U.S. Arms Control and Disarmament Agency, *Documents on Disarmament, 1969* (Washington, D.C.: Government Printing Office, 1970), pp. 268-269.

17. Friedman, *Law of War*, pp. 229, 318; "Protocol Additional to the Geneva Conventions of 12 August 1949, and Relating to the Protection of Victims of International Arms Conflicts (Protocol I)," mimeo. (Geneva: Diplomatic Conference on the Reaffirmation and Development of International Humanitarian Law Applicable to Armed Conflicts, 1977), p. 25.

18. Falk, *Vietnam War*, p. 209.

19. McDougal and Feliciano, *Law*, p. 72.

20. Robert W. Tucker, "The Law of War and Neutrality at Sea," *International Law Studies* (1955):48, n. 8.

21. Tucker himself recognizes this fact and is resigned to it: "The general principles of the law of war have always suffered under certain limitations which have served to limit their potential utility. The very character of these general principles must lead to difficulties of interpretation and application. These difficulties are magnified, of course, by the fact that the principal subjects of the law normally must interpret and apply the law. The consequences of this latter condition admittedly are not without effect upon the whole of the law of war. No rule can be so specific that its interpretation and application remain unaffected by the condition of extreme decentralization characteristic of international law. Nevertheless, a measure of certainty may at least be achieved to the degree that the general principles of the law of war are given a more concrete form through the establishment of detailed rules of custom and—particularly—convention. In the absence of such detailed regulation their interpretation and application with respects to the rapidly changing weapons and methods of warfare will be—almost of necessity—a matter of endless controversy and uncertainty." Ibid., p. 50.

22. McDougal and Feliciano, *Law*, p. 654. See also Sir Charles Webster and Noble Frankland, *The Strategic Air Offensive Against Germany, 1939-1945* (London: Her Majesty's Stationery Office, 1961), 1:301, 2:115.

23. Thomas and Thomas, *Legal Limits*, p. 203.

24. Richard A. Falk, "The Claimants of Hiroshima," in Richard A. Falk and Saul H. Mendlovitz, *The Strategy of World Order* (New York: World Law Fund, 1966), 1:242-244.

25. U.S. Arms Control and Disarmament Agency, *Documents on Disarmament, 1961* (Washington, D.C.: Government Printing Office, 1962), pp. 648-650.

26. U.S. Arms Control and Disarmament Agency, *Documents on Disarmament, 1972* (Washington, D.C.: Government Printing Office, 1974), p. 831.

27. Leland Goodrich and Edvard Hambro, *Charter of the United Nations* (New York: Columbia University Press, 1969), p. 345.

28. McDougal and Feliciano, *Law*, p. 617.

29. Ibid., pp. 617-671.

30. Exceptions include the use of gas by the Japanese in China during World War II and by the Egyptians in Yemen during the 1960s. For details, see Thomas and Thomas, *Legal Limits*, p. 203.

31. Ibid., p. 145.

32. Ibid., p. 185. For a similar assessment, see Frederic J. Brown, *Chemical Warfare: A Study of Restraints* (Princeton, N.J.: Princeton University Press, 1968).

33. "Protocol Additional."

34. Ibid., pp. 38-39.

35. Ibid., p. 37.

36. Ibid.

37. Ibid., p. 38.

38. For the treaty text, see Stockholm International Peace Research Institute, *World Armaments and Disarmament: SIPRI Yearbook 1978* (London: Taylor and Francis, 1978), pp. 392-397.

39. Hans Morgenthau, *Politics among Nations*, 4th ed. (New York: Alfred A. Knopf, 1976), pp. 355-356, and Robert Osgood and Robert W. Tucker, *Force, Order and Justice* (Baltimore: Johns Hopkins University Press, 1967), pp. 41-120.

40. McDougal and Feliciano, *Law*, p. 616.

41. See U.S. Arms Control and Disarmament Agency, *Documents on Disarmament, 1969* (Washington, D.C.: Government Printing Office, 1970), pp. 617-618.

42. Ibid., p. 712.

43. U.S. Arms Control and Disarmament Agency, *Documents on Disarmament, 1970* (Washington, D.C.: Government Printing Office, 1971), p. 504.

44. Ibid., pp. 308-309.

45. Reviews of these events and those through 1978 can be found in Stockholm International Research Institute, *World Armaments*, pp. 382-390; U.S. General Assembly, *Report of the Conference on the Committee on Disarmament*, General Assembly Official Records, 32 Session, Supplement No. 27, A/32/27 (New York: United Nations, 1977), 1:62-69; U.N. General Assembly, *Special Report of the Committe on Disarmament*, General Assembly Official Records, 10 Special Session Supplement, No. 2, A/S-10/2 (New York: United Nations, 1978), 2:38-44.

46. U.S. Arms Control and Disarmament Agency, "Statement Delivered by Dr. Fred C. Iklé, Director, U.S. Arms Control and Disarmament Agency, 'United Nations General Assembly, November 18, 1976,' " *Press Release* (Washington, D.C.: U.S. Arms Control and Disarmament Agency, November 18, 1978), p. 10.

47. "Agreed Joint US-USSR Proposal on Major Elements of a Treaty Prohibiting the Development, Production, Stockpiling and Use of Radiological Weapons," mimeo. (1979).

48. Ibid., art. II.

49. Falk, "Claimants," p. 297.

50. Ibid., p. 300.

51. Ibid., pp. 300-302.

52. For elaboration, see Richard Ullman, "No First Use of Nuclear Weapons," *Foreign Affairs* 50 (July 1972):669-683.

53. For elaboration see ibid.; Robert C. Tucker, "No First Use of Nuclear Weapons: A Proposal," and Richard A. Falk, "Some Thoughts in Support of a No-First-Use Proposal," in Robert C. Tucker, et al. *Proposals for No First Use of Nuclear Weapons: Pros and Cons, Policy Memorandum No. 28* (Princeton, N.J.: Center of International Studies, Princeton University, 1963), pp. 1-19, 37-56, respectively.

54. Tucker, "No First Use," p. 2.

55. For an assessment of bargaining strategies at the outset of negotiation, see Bennett Ramberg, "Tactical Advantages of Opening Positioning Strategies: Lessons from the Seabed Arms Control Talks, 1967-1970," *Journal of Conflict Resolution* 21 (December 1977):685-700.

56. Abram Chayes, "An Inquiry into the Workings of Arms Control Agreements," *Harvard Law Review* 135 (March 1972):907, 932.

57. Ibid., pp. 935-936.

58. Falk, "Some Thoughts in Support," pp. 38-39.

59. Brown, *Chemical Warfare*, pp. 293-294.

60. This article draws on art. III, paragraph 5, of the *Treaty on the Prohibition of the Emplacement of Nuclear Weapons and Other Weapons of Mass Destruction on the Seabed and the Ocean Floor and in the Subsoil Thereof* (hereafter cited as *Seabed Treaty*) and on art. V of the *Convention*

on the Prohibition of the Development, Production and Stockpiling of Bacteriological (Biological) and Toxin Weapons and on Their Destruction (hereafter cited as *Bacteriological Treaty*), in Arms Control and Disarmament Agency, *Arms Control and Disarmament Agreements*, pp. 97, 119.

61. This article draws on art. 56, paragraph 7, "Protocol Additional," p. 39.

62. Art. IV, paragraph 1, of *Treaty on the Non-Proliferation of Nuclear Weapons*, in Arms Control and Disarmament Agency, *Arms Control and Disarmament Agreements*, p. 87.

63. Art. VI, *Seabed Treaty*, p. 98.

64. This article draws on ibid. and *Bacteriological Treaty*, pp. 120-121.

65. *Seabed Treaty*, pp. 98-99.

66. Ibid., p. 99.

67. "Protocol Additional," annex I, chap. 6, art. 16, p. 88.

68. Augustin M. Prentiss, *Civil Defense in Modern War* (New York: McGraw-Hill, 1951), p. 168. The active-passive civil-defense distinction appears in Harold Brown, *Department of Defense Annual Report Fiscal Year 1980* (Washington, D.C.: Department of Defense, 1980), p. 69.

69. Richard L. Garwin, "Effective Military Technology for the 1980s," *International Security* 1 (Fall 1976):53-56.

70. Curtis E. Harvey, "Civil Defense Abroad in Review," in Eugene P. Wigner, *Survival and the Bomb: Methods of Civil Defense* (Bloomington: Indiana University Press, 1969), pp. 159-162, and Walter H. Murphey and Bjorn Klinge, "Civil Defense Abroad," in Eugene P. Wigner, *Who Speaks for Civil Defense* (New York: Charles Scribner's, 1968), pp. 76-80.

71. Harvey, "Civil Defense Abroad," pp. 156-168; Swedish Government Committee on Radioactive Waste, *Spent Nuclear Fuel and Radioactive Waste*, SOU 1976:32 (Stockholm: Swedish Government Committee on Radioactive Waste, 1976), pp. 64-65.

72. Harvey, "Civil Defense," pp. 149-177; and Murphey and Kling, "Civil Defense," pp. 75-86.

73. Nuclear Regulatory Commission, *Regulatory Guide 1.101: Emergency Planning for Nuclear Power Plants* (Washington, D.C.: Nuclear Regulatory Commission, March 1977, Revision 1).

74. American Physical Society Study Group on Light-Water Reactor Safety, *Review of Modern Physics* 47, supplement no. 1 (Summer 1975):S52. Further elaboration will be found in *Civil Defense Aspects of the Three Mile Island Nuclear Accident*, Hearings, 96th Cong., 1st sess., 1979.

75. President's Commission on the Accident at Three Mile Island, *The Accident at Three Mile Island* (Washington, D.C.: President's Commission on Three Mile Island, 1979), p. 73. Studies that examine the marine option include George Yadigaroglu and Stephen O. Andersen, "Novel Siting Solutions for Nuclear Power Plants," *Nuclear Safety* 15 (November-December

1974):654-657, and Otto H. Klepper and Truman D. Anderson, "Siting Considerations for Future Offshore Nuclear Energy Stations," *Nuclear Technology* 22 (May 1974):160-169.

76. Donald McIsaac, ed., *United States Strategic Bombing Survey* (New York: Garland Publishing Co., 1976), 6:31.

77. Ibid., p. 54.

78. Ibid., p. 55.

79. M.B. Watson, et al., *Underground Nuclear Power Plant Siting* (Pasadena: Environmental Quality Laboratory, California Institute of Technology, 1972), appendix 1-2.

80. *Nucleonics Weeks*, July 28, 1977, p. 2.

81. Conrad V. Chester and Rowena O. Chester, "Civil Defense Implications of the U.S. Nuclear Power Industry During a Large Nuclear War in the Year 2000," *Nuclear Technology* 31 (December 1976):329.

82. J.H. Crowley, et al., "Underground Nuclear Plant Siting: A Technical and Safety Assessment," *Nuclear Safety* 15 (September-October 1974):531.

83. Ibid.

84. *Nucleonics Week*, July 28, 1977.

85. Crowley, et al., "Underground Nuclear Plant Siting."

86. Thomas C. Hollocher, "Storage and Disposal of High Level Radioactive Wastes," in Union of Concerned Scientists, *The Nuclear Fuel Cycle*, rev. ed. (Cambridge: MIT Press, 1975), pp. 267-269.

87. Private communication, April 9, 1980. Chester supports the contention that water may be a significant container of radionuclides by citing in addition to the Three Mile Island accident the destruction of three nuclear propulsion reactors on board the American submarines Thresher and Scorpion and at least one Soviet atom-powered vessel, which sank due to non-nuclear causes without spreading "significant" contamination. Chester also notes in his letter that most random damage by conventional munitions will result in release of water in the containment. This would help reduce the consequences of willful destruction. He argues that for the worst consequences to occur one needs an almost dry containment.

88. David Okrent, et al., *Post-Accident Filtration as a Means of Improving Containment Effectiveness* (Los Angeles: University of California, Los Angeles School of Engineering and Applied Science, 1977).

89. Ibid., pp. 1-2.

90. Theodore B. Taylor, "Reactor Safety Considerations Related to Sabotage and Wartime Bombardment of Nuclear Power Plants" (unpublished manuscript, 1968), pp. 11-13. Conrad Chester believes that the safest design is the molton suit breeder reactor. Private communication, April 9, 1980.

91. J.T. Rogers, "Candu Moderator Provides Ultimate Heat Sink in a Loca," *Nuclear Engineering International* 24 (January 1979):38-41.

92. Mason Willrich and Theodore B. Taylor, *Nuclear Theft: Risks and Safeguards* (Cambridge, Mass.: Ballinger, 1974), p. 42.

93. Richard E. Webb, *The Accident Hazards of Nuclear Power Plants* (Amherst: University of Massachusetts Press, 1976), pp. 152-153; A.W. Barsell "An Assessment of HTGR Accident Consequences," *Nuclear Safety* 18 (November-December 1977):761-773; and G.L. Wessman and T.R. Moffette, "Safety Design Bases of the HTGR," *Nuclear Safety* 14 (November-December 1973):618-634.

94. K.G. Beckhurts, et al., "The Gas Cooled High Temperature Reactor: Perspectives, Problems and Programs," IAEA-CN-36/94 (Vienna: International Atomic Energy Agency, 1977), pp. 1-2.

95. Office of Technology Assessment, *Nuclear Proliferation and Safeguards* (Washington, D.C.: Office of Technology Assessment, 1977), appendix, vol. 2, pt. I, p. V-173.

96. Hollacher, "Storage and Disposal," pp. 238-244, 267-269.

97. Ibid., pp. 246-247, and National Academy of Engineering and National Academy of Sciences, *Solidification of High Level Radioactive Wastes*, NUREG/CR-0895 (Washington, D.C.: U.S. Nuclear Regulatory Commission, July 1979), pp. 18-26.

98. This criterion is suggested by E. Linn Draper, Jr., "A Rational Energy Strategy: One Man's Opinion," in U.S. Senate, Select Committee on Small Business and the Committee on Interior and Insular Affairs, *Alternative Long-Range Energy Strategies* Hearings, 94th Cong., 2d sess., pp. 298-300.

99. Amory B. Lovins, "Energy Strategy: The Road Not Taken," *Foreign Affairs* 105 (October 1976):65-96. Reprinted by permission from *Foreign Affairs*, October 1976. Copyright 1976 by Council on Foreign Relations, Inc.

100. Amory Lovins, "Re-Examining the Nature of the ECE Energy Problem" (Los Angeles: California Seminar on Arms Control and Foreign Policy), pp. 4-6.

101. Lovins, "Energy Strategy," p. 73.

102. Ibid., pp. 77-78.

103. Ibid., p. 77.

104. Ibid.

105. Senate, Select Committee on Small Business and the Committee on Interior and Insular Affairs, *Alternative Long-Range Energy Strategies*.

106. Ibid., pp. 405-413.

107. Sec. 123, *Atomic Energy Act of 1954* (as amended) in Congressional Research Service, *Nuclear Proliferation Factbook* (Washington, D.C.: Government Printing Office, 1977), pp. 34-35.

108. Energy Research and Development Administration, *Final Environmental Statement: U.S. Nuclear Power Export Activities* (Springfield, Va.: National Technical Information Service, 1976), p. 1-14.

109. U.S. Senate, Committee on Government Operations, *Export Reorganization Act of 1976*, Hearings, 94th Cong., 2d sess., 1976, p. 885.

110. Ibid., pp. 886-887.

111. Ibid., p. 886.

112. Energy Research and Development Administration, *Final Environmental Statement*, pp. i, 1-1.

113. Ibid., p. i.

114. Ibid., p. 1-21.

115. Ibid., sec. 4.

116. Ibid., sec. 5-8.

117. Ibid, p. iii.

118. Ibid., p. 14-12.

119. U.S. Congress, House, Committee on Foreign Affairs, Subcommittees on International Organizations and Movements and on the Near East and South Asia, *Hearings on U.S. Foreign Policy and the Export of Nuclear Technology to the Middle East*, 92d Cong., 2d sess., 1974, p. 93.

120. Ibid.

121. Ibid., pp. 188, 198.

122. Ibid., pp. 100-101.

123. Congressional Research Service, *Analysis of Arms Control Impact Statements Submitted in Connection with the Fiscal Year 1976 Budget Request* (Washington, D.C.: Government Printing Office, 1977), pp. 350-351.

124. Ibid.

125. Ibid., pp. 4-5.

126. Ibid., pp. 6-7.

127. Ibid., pp. 8-11.

128. *Treaty Establishing the European Atomic Energy Community*, art. 77.

129. Telephone interview with the staff of the European Information Center, Washington, D.C., 1977.

130. For a review of Cocom's and Chicom's development, see Gunner Adler-Karlson, *Western European Warfare, 1947-1966* (Stockholm: Almqvist and Wiksell, 1968), and Robert E. Klitgaard, *National Security and Export Controls*, R-1432-1-ARPA/CIEP (Santa Monica, Calif.: Rand Corporation, 1974).

131. Adler-Karlson, *Western European Warfare*, pp. 54, 95.

132. Ibid., pp. 187-200, and Klitgaard, *National Security*, p. 71.

133. John A. King, Jr., *Economic Development Projects and Their Appraisal* (Baltimore: Johns Hopkins University Press, 1976), p. 5.

134. Ibid., pp. 3-15.

135. Edward S. Mason and Robert E. Asher, *The World Bank Since Bretton Woods* (Washington, D.C.: Brookings Institution, 1973), pp. 257-259.

136. Nuclear Suppliers Group, "Nuclear Suppliers Group Guidelines for Nuclear Transfers," mimeo. (1977).

137. Stockholm International Peace Research Institute, *World Armaments and Disarmament: SIPRI Yearbook 1977* (Cambridge, Mass.: MIT Press, 1977), p. 22.

138. Stockholm International Peace Research Institute, *World Armaments and Disarmament: SIPRI Yearbook 1978*, p. 28.

139. Wojciech Morawiecki, "The IAEA's Role in Promoting Physical Protection of Nuclear Material and Facilities," *International Atomic Energy Agency Bulletin* 20 (June 1978):39-45.

Index

Afghanistan, 97, Soviet use of gas in, 118, as a threat to Iran, 102

Africa. *See* names of specific African countries

Argentina: nuclear energy program, 14, 16

Arms Control and Disarmament Act (as amended 1975), 149-151

Arms Control and Disarmament Agency, 145, 155; assessment of U.S. weapons acquisitions; evaluation of U.S. nuclear exports, 146, 148-149

Asia, 90-103. *See also* names of specific Asian countries

Atom bomb survivors: physical effects of radiation on, 5, 7; psychological and sociological effects of radiation on, 9-10, 69

Austria: nuclear energy program 14, 16

Barnwell, South Carolina reprocessing plant, 57

Barsebäck reactor: biological consequences from destruction of, 46-51; contamination from destruction of, 46-48, 69

Belgium: nuclear energy program, 14, 16

Beyea, Jan, 34, 46, 48, 50, 60, 78, 82, 110, 134

Biological consequences of irradiation. *See* Radiation, biological impact

Boiling water reactor (BWR): consequences of operational failures, 28-29, 32-33; illustrations of, 21, 24; operation of, 20; safety features of, 22, 24-25. *See also* Barsebäck reactor

Brazil: nuclear energy program, 14, 16

Brown, Frederic, 127

Brown, Harold, 175, 176

Bulgaria: nuclear energy program

Canada: Cocom participation of, 151; nuclear energy program, 14, 16, 140; nuclear supplier participation of, 153

Candu reactor, 140

Chayes, Abram, 126-127

Chester, Conrad, 173, 184

Chester, Conrad and Rowena Chester, 61

Chile: nuclear energy program, 16

China, Peoples Republic of: and Korean War, 15; nuclear energy program, 91; nuclear weapons use policy, 125; strategic implications of nuclear energy vulnerability in war for, 18, 78-80, 91, 94, 97, 109; strategic implications of Soviet nuclear energy vulnerability in war for, 78-80

China, Republic of. *See* Taiwan

Civil defense, 132-134, 163; active 132, 183; evacuation as, 35-36, 38-39, 42-43, 133-134; in France, 133; passive, 132-134, 183; sheltering as, 36, 133-134; in Soviet Union, 78, 133; in Sweden, 134; in Switzerland, 134

Coercive diplomacy and nuclear energy, 72-73, 109, 162; applied to China, 91, 94; Cooper on, 71-72; Europe, 80, 81; Korea, 91; Middle East, 88, 90; Schelling on, 72; South Africa, 104-106; South Asia, 97; Soviet Union, 78-80; United States, 106; West Asia, 103

Conference of the Committee on Disarmament (CCD), 114, 122, 123

Congressional Research Service (CRS): review of Arms Control Impact Statements, 150

Consultative Committee (Cocom), 145, 151-152

Contamination resulting from nuclear energy destruction, xvi, 25-29, 30-69, 161, 162; consequences in Europe, 80-82, 109, 110; Korea, 90-91, 110; Middle East, 88, 109-110; South Africa, 104; Soviet Union, 78, 109-110; United States, 106-108; West Asia, 103-104

About the Author

Bennett Ramberg is a research fellow at the University of California at Los Angeles Center for International and Strategic Affairs. A graduate of the University of Southern California, he received the M.A. and Ph.D. from The Johns Hopkins University School of Advanced International Studies. Dr. Ramberg has been a research fellow at Stanford University's Arms Control Program and Princeton University's Center of International Studies. He also has been a congressional consultant. His publications include *The Seabed Arms Control Negotiations: A Study of Multilateral Arms Control Conference Diplomacy* (1978); "Destruction of Nuclear Energy Facilities in War: A Proposal for Legal Restraint," *World Order Studies Occasional Paper No. 7* (1978); and "Tactical Advantages of Opening Positioning Strategies: Lessons from the Seabed Arms Control Talks, 1967-1970," *Journal of Conflict Resolution* (1977).

Center of International Studies: List of Publications

Gabriel A. Almond, *The Appeals of Communism* (Princeton University Press, 1954).

William W. Kaufmann, ed., *Military Policy and National Security* (Princeton University Press, 1956).

Klaus Knorr, *The War Potential of Nations* (Princeton University Press, 1956).

Lucien W. Pye, *Guerrilla Communism in Malaya* (Princeton University Press, 1956).

Charles De Visscher, *Theory and Reality in Public International Law*, trans. by P.E. Corbett (Princeton University Press, 1957; rev. ed., 1968).

Bernard C. Cohen, *The Political Process and Foreign Policy: The Making of the Japanese Peace Settlement* (Princeton University Press, 1957).

Myron Weiner, *Party Politics in India: The Development of a Multi-Party System* (Princeton University Press, 1957).

Percy E. Corbett, *Law in Diplomacy* (Princeton University Press, 1959).

Rolf Sannwald and Jacques Stohler, *Economic Integration: Theoretical Assumptions and Consequences of European Unification*, trans. by Herman Karreman (Princeton University Press, 1959).

Klaus Knorr, ed., *NATO and American Security* (Princeton University Press, 1959).

Gabriel A. Almond and James S. Coleman, eds., *The Politics of the Developing Areas* (Princeton University Press, 1960).

Herman Kahn, On Thermonuclear War (Princeton University Press, 1960).

Sidney Verba, *Small Groups and Political Behavior: A Study of Leadership* (Princeton University Press, 1961).

Robert J.C. Butow, *Tojo and the Coming of the War* (Princeton University Press, 1961).

Glenn H. Snyder, *Deterrence and Defense: Toward a Theory of National Security* (Princeton University Press, 1961).

Klaus Knorr and Sidney Verba, eds., *The International System: Theoretical Essays* (Princeton University Press, 1961).

Peter Paret and John W. Shy, *Guerrillas in the 1960's* (Praeger, 1962).

George Modelski, *A Theory of Foreign Policy* (Praeger, 1962).

Klaus Knorr and Thornton Read, eds., *Limited Strategic War* (Praeger, 1963).

Frederick S. Dunn, *Peace-Making and the Settlement with Japan* (Princeton University Press, 1963).

197

Arthur L. Burns and Nina Heathcote, *Peace-Keeping by United Nations Forces* (Praeger, 1963).

Richard A. Falk, *Law, Morality, and War in the Contemporary World* (Praeger, 1963).

James N. Rosenau, *National Leadership and Foreign policy: A Case Study in the Mobilization of Public Support* (Princeton University Press, 1963).

Gabriel A. Almond and Sidney Verba, *The Civic Culture: Political Attitudes and Democracy in Five Nations* (Princeton University Press, 1963).

Bernard C. Cohen, *The Press and Foreign Policy* (Princeton University Press, 1963).

Richard L. Sklar, *Nigerian Political Parties: Power in an Emergent African Nation* (Princeton University Press, 1963).

Peter Paret, *French Revolutionary Warfare from Indochina to Algeria: The Analysis of a Political and Military Doctrine* (Praeger, 1964).

Harry Eckstein, ed., *Internal War: Problems and Approaches* (Free Press, 1964).

Cyril E. Black and Thomas P. Thornton, eds., *Communism and Revolution: The Strategic Uses of Political Violence* (Princeton University Press, 1964).

Miriam Camps, *Britain and the European Community 1955-1963* (Princeton University Press, 1964).

Thomas P. Thornton, ed., *The Third World in Soviet Perspective: Studies by Soviet Writers on the Developing Areas* (Princeton University Press, 1964).

James N. Rosenau, ed., *International Aspects of Civil Strife* (Princeton University Press, 1964).

Sidney I. Ploss, *Conflict and Decision-Making in Soviet Russia: A Case Study of Agricultural Policy, 1953-1963* (Princeton University Press, 1965).

Richard A. Falk and Richard J. Barnet, eds., *Security in Disarmament* (Princeton University Press, 1965).

Karl von Vorys, *Political Development in Pakistan* (Princeton University Press, 1965).

Harold and Margaret Sprout, *The Ecological Perspective on Human Affairs, with Special Reference to International Politics* (Princeton University Press, 1965).

Klaus Knorr, *On the Uses of Military Power in the Nuclear Age* (Princeton University Press, 1966).

Harry Eckstein, *Division and Cohesion in Democracy: A Study of Norway* (Princeton University Press, 1966).

Cyril E. Black, *The Dynamics of Modernization: A Study in Comparative History* (Harper and Row, 1966).

Peter Kunstadter, ed., *Southeast Asian Tribes, Minorities, and Nations* (Princeton University Press, 1967).

E. Victor Wolfenstein, *The Revolutionary Personality: Lenin, Trotsky, Gandhi* (Princeton University Press, 1967).

Leon Gordenker, *The UN Secretary-General and the Maintenance of Peace* (Columbia University Press, 1967).

Oran R. Young, *The Intermediaries: Third Parties in International Crises* (Princeton University Press, 1967).

James N. Rosenau, ed., *Domestic Sources of Foreign Policy* (Free Press, 1967).

Richard F. Hamilton, *Affluence and the French Worker in the Fourth Republic* (Princeton University Press, 1967).

Linda B. Miller, *World Order and Local Disorder: The United Nations and Internal Conflicts* (Princeton University Press, 1967).

Henry Bienen, *Tansania: Party Transformation and Economic Development* (Princeton University Press, 1967).

Wolfram F. Hanrieder, *West German Foreign Policy, 1949-1963: International Pressures and Domestic Response* (Stanford University Press, 1967).

Richard H. Ullman, *Britain and the Russian Civil War: November 1918-Febrary 1920* (Princeton University Press, 1968).

Robert Gilpin, *France in the Age of the Scientific State* (Princeton University Press, 1968).

William B. Bader, *The United States and the Spread of Nuclear Weapons* (Pegasus, 1968).

Richard A. Falk, *Legal Order in a Violent World* (Princeton University Press, 1968).

Cyril E. Black, Richard A. Falk, Klaus Knorr and Oran R. Young, *Neutralization and World Politics* (Princeton University Press, 1968).

Oran R. Young, *The Politics of Force: Bargaining During International Crises* (Princeton University Press, 1969).

Klaus Knorr and James N. Rosenau, eds., *Contending Approaches to International Politics* (Princeton University Press, 1969).

James N. Rosenau, ed., *Linkage Politics: Essays on the Convergence of National and International Systems* (Free Press, 1969).

John T. McAlister, Jr., *Viet Nam: The Origins of Revolution* (Knopf, 1969).

Jean Edward Smith, *Germany Beyond the Wall: People, Politics and Prosperity* (Little, Brown, 1969).

James Barros, *Betrayal from Within: Joseph Avenol, Secretary-General of the League of Nations, 1933-1940* (Yale University Press, 1969).

Charles Hermann, *Crises in Foreign Policy: A Simulation Analysis* (Bobbs-Merrill, 1969).

Robert C. Tucker, *The Marxian Revolutionary Idea: Essays on Marxist Thought and Its Impact on Radical Movements* (W.W. Norton, 1969).

Harvey Waterman, *Political Change in Contemporary France: The Politics of an Industrial Democracy* (Charles E. Merrill, 1969).

Cyril E. Black and Richard A. Falk, eds., *The Future of the International Legal Order*. Vol. I: *Trends and Patterns* (Princeton University Press, 1969).

Ted Robert Gurr, *Why Men Rebel* (Princeton University Press, 1969).

C. Sylvester Whitaker, *The Politics of Tradition: Continuity and Change in Northern Nigeria 1946-1966* (Princeton Unversity Press, 1970).

Richard A. Falk, *The Status of Law in International Society* (Princeton University Press, 1970).

John T. McAlister, Jr. and Paul Mus, *The Vietnamese and Their Revolution* (Harper & Row, 1970).

Klaus Knorr, *Military Power and Potential* (D.C. Heath, 1970).

Cyril E. Black and Richard A. Falk, eds., *The Future of the International Legal Order*. Vol. II: *Wealth and Resources* (Princeton University Press, 1970).

Leon Gordenker, ed., *The United Nations in International Politics* (Princeton University Press, 1971).

Cyril E. Black and Richard A. Falk, eds., *The Future of the International Legal Order*. Vol. III: *Conflict Management* (Princeton University Press, 1971).

Francine R. Frankel, *India's Green Revolution: Economic Gains and Political Costs* (Princeton University Press, 1971).

Harold and Margaret Sprout, *Toward a Politics of the Planet Earth* (Van Nostrand Reinhold Co., 1971).

Cyril E. Black and Richard A. Falk, eds., *The Future of the International Legal Order*. Vol. IV: *The Structure of the International Environment* (Princeton University Press, 1972).

Gerald Garvey, *Energy, Ecology, Economy* (W.W. Norton, 1972).

Richard H. Ullman, *The Anglo-Soviet Accord* (Princeton University Press, 1973).

Klaus Knorr, *Power and Wealth: The Political Economy of International Power* (Basic Books, 1973).

Anton Bebler, *Military Rule in Africa: Dahomey, Ghana, Sierra Leone, and Mali* (Praeger Publishers, 1973).

Robert C. Tucker, *Stalin as Revolutionary 1879-1929: A Study in History and Personality* (W.W. Norton, 1973).

Edward L. Morse, *Foreign Policy and Interdependence in Gaullist France* (Princeton University Press, 1973).

Henry Bienen, *Kenya: The Politics of Participation and Control* (Princeton University Press, 1974).

Gregory J. Massell, *The Surrogate Proletariat: Moslem Women and Revolutionary Strategies in Soviet Central Asia, 1919-1929* (Princeton University Press, 1974).

James N. Rosenau, *Citizenship Between Elections: An Inquiry Into The Mobilizable American* (Free Press, 1974).

Ervin Laszlo, *A Strategy for the Future: The Systems Approach to World Order* (George Braziller, 1974).

R.J. Vincent, *Nonintervention and International Order* (Princeton University Press, 1974).

Jan H. Kalicki, *The Pattern of Sino-American Crises: Political-Military Interactions in the 1950s* (Cambridge University Press, 1975).

Klaus Knorr, *The Power of Nations: The Political Economy of International Relations* (Basic books, Inc., 1975).

James P. Sewell, *UNESCO and World Politics: Engaging in International Relations* (Princeton University Press, 1975).

Richard A. Falk, *A Global Approach to National Policy* (Harvard University Press, 1975).

Harry Eckstein and Ted Robert Gurr, *Patterns of Authority: A Structural Basis for Political Inquiry* (John Wiley & Sons, 1975).

Cyril E. Black, Marius B. Jansen, Herbert S. Levine, Marion J. Levy, Jr., Henry Rosovsky, Gilbert Rozman, Henry D. Smith, II, and S. Frederick Starr, *The Modernization of Japan and Russia* (Free Press, 1975).

Leon Gordenker, *International Aid and National Decisions: Development Programs in Malawi, Tanzania, and Zambia* (Princeton University Press, 1976).

Carl von Clausewitz, *On War*, edited and translated by Michael Howard and Peter Paret (Princeton University Press, 1976).

Gerald Garvey and Lou Ann Garvey, *International Resource Flows* (Lexington Books, D.C. Heath, 1977).

Walter F. Murphy and Joseph Tanenhaus, *Comparative Constitutional Law: Cases and Commentaries* (St. Martin's Press, 1977).

Gerald Garvey, *Nuclear Power and Social Planning: The City of the Second Sun* (Lexington Books, D.C. Heath, 1977).

Richard E. Bissell, *Apartheid and International Organizations* (Westview Press, 1977).

David P. Forsythe, *Humanitarian Politics: The International Committee of the Red Cross* (John Hopkins University Press, 1977).

Paul E. Sigmund, *The Overthrow of Allende and the Politics of Chile, 1964-1976* (University of Pittsburgh Press, 1977).

Henry S. Bienen, *Armies and Parties in Africa* (Holmes and Meier 1978).

Harold and Margaret Sprout, *The Context of Environmental Politics: Unfinished Business for America's Third Century* (University Press of Kentucky, 1978).

Samuel S. Kim, *China, The United Nations, and World Order* (Princeton University Press, 1979).

S. Basheer Ahmed, *Nuclear Fuel and Energy* (Lexington Books, D.C. Heath, 1979).

Robert Johansen, *The National Interest and the Human Interest: An Analysis of U.S. Foreign Policy* (Princeton University Press, 1980).

Center for International and Strategic Affairs University of California, Los Angeles Series Studies in International and Strategic Affairs: List of Publications

William Potter, ed., *Verification and SALT: The Challenge of Strategic Deception* (Westview Press, 1980).

Paul Jabber, *Not by War Alone: The Politics of Arms Control in the Middle East* (University of California Press, 1980).

Roman Kolkowicz and Andrej Korbonski, eds., *Civil-Military Relations in Socialist and Modernizing Societies* (Allan and Unwin, forthcoming).

Steven Spiegel, ed., *The Middle East and the Western Alliance* (Allen and Unwin, forthcoming).

Bernard Brodie, Michael Intriligator, and Roman Kolkowicz, eds., *National Security and International Stability* (forthcoming).